1973   BENJAMIN F. FAIRLESS MEMORIAL LECTURES

©1974 CARNEGIE-MELLON UNIVERSITY

Library of Congress Catalog Card Number 74-77475
ISBN Number 0-231-03864-X

The Productive Society

Patrick E. Haggerty

Distributed by Columbia University Press

New York-London

The Benjamin F. Fairless Memorial Lectures endowment fund has been established at Carnegie-Mellon University to support an annual series of lectures. An internationally known figure from the worlds of business, government, or education is invited each year to present three lectures at Carnegie under the auspices of its Graduate School of Industrial Administration. In general, the lectures will be concerned with some aspects of business or public administration; the relationships between business and government, management and labor; or a subject related to the themes of preserving economic freedom, human liberty, and the strengthening of individual enterprise—all of which were matters of deep concern to Mr. Fairless throughout his career.

Mr. Fairless was president of United States Steel Corporation for fifteen years, and chairman of the board from 1952 until his retirement in 1955. A friend of Carnegie-Mellon University for many years, he served on the board of trustees from 1952 until his death. In 1959 he was named honorary chairman of the board. He was also a leader and co-chairman of Carnegie-Mellon's first development program, from its beginning in 1957.

Patrick E. Haggerty is chairman of the board and chief corporate officer of Texas Instruments Incorporated and has been affiliated with TI and its predecessor company in various managerial capacities since 1945.

He holds a B.S.E.E. from Marquette University, and his contributions have been recognized with honorary doctorates from eight universities. Some additional honors include the Founders Award, Institute of Electrical and Electronics Engineers; Medalist, Industrial Research Institute; Eminent Membership, Eta Kappa Nu; John Fritz Medal (awarded collectively by the following engineering societies: ASCE, AIME, ASME, IEEE, AIChE); Marquette University Alumnus of the Year.

His technical-managerial-industrial skills also are utilized in various other activities. At present, he serves as chairman of the National Council on Educational Research (policy-making and advisory committee to the National Institute of Education), as a member of the Boards of Directors and Executive Committees of Rockefeller University and the University of Dallas, the Board of Directors of A. H. Belo Corporation, the International Advisory Committee of Chase Manhattan Bank, the Commission and Executive Committee of the Trilateral Commission, and as a member of the Business Council and the National Academy of Engineering. Prior service has included being a Governor of the U.S. Postal Service, Members and Funds Chairman of the American National Red Cross, and a member of the Business Advisory Committee, Graduate School of Industrial Administration of Carnegie-Mellon University.

Mr. Haggerty was co-chairman of the committee which effected the merger of the Institute of Radio Engineers and the American Institute of Electrical Engineers. The merger brought about the Institute of Electrical and Electronics Engineers, the world's largest professional society, and brought together a logical spectrum of technologies and engineering capabilities to serve society and the profession more fully in the years ahead. He is a Fellow of the IEEE.

Mr. Haggerty has had a long interest in studying and improving productivity in today's society.

Introduction

The theme of these lectures has been an abiding interest of mine over a number of years. In particular, three of my previous papers relate rather specifically to this present discussion. These are:

Technology and the American Standard of Living
   Charles M. Schwab Memorial Lecture
   May 27, 1970

The Productive Society
   Charles P. Steinmetz Memorial Lecture
   May 12, 1971 (subsequently published in the September 1971 *Spectrum*)

Industrial Research and Development
   Rockefeller University Seminar on Science and the
   Evolution of Public Policy,
   March 16, 1972 (subsequently published in the book
   of the same name, edited by Dr. James A. Shannon)

I have drawn from these three papers wherever it seemed appropriate and in some detail, but I have not attempted to reference them at each such point.

In his lectures last year, Mr. Donald C. Burnham discussed *Productivity Improvement* with great clarity and conviction. I am honored to join him and the other distinguished men who have participated in these Fairless Lectures. Although there are inevitably a few overlaps, I have concentrated on the overall societal systems requisite to high productivity, and I would hope, therefore, that Mr. Burnham's presentation and mine are suitably complementary.

The lectures, as delivered, were based on an abbreviated form of the written version that follows. Chapter I was the basis for Lecture I; Chapter II for Lecture II; and Chapters III and IV for Lecture III.

1973

The Productive Society

eleven

The Productive Society

I. Is Growth Obsolete?

I. Is Growth Obsolete?

A State, I said, arises, as I conceive, out of the needs of mankind; no one is self-sufficing, but all of us have many wants. Can any other origin of a State be imagined?

There can be no other.

Then, as we have many wants, and many persons are needed to supply them, one takes a helper for one purpose and another for another; and when these partners and helpers are gathered together in one habitation the body of inhabitants is termed a State.

True, he said.

And they exchange with one another, and one gives, and another receives, under the idea that the exchange will be for their good.

Very true.

Then, I said, let us begin and create in idea a State; and yet the true creator is necessity, who is the mother of our invention.[1]

That, of course, is a quotation from Plato's "Discourses on the Republic," and it is one instance as early as 400 B.C. of man's preoccupation with the design of modes of law and function to establish a more perfect society.

The impelling motives behind one such design hardly could have been stated better than in these words:

We the People of the United States, in Order to form a more perfect Union, establish Justice, ensure domestic Tranquillity, provide for the common defense, promote the general Welfare, and secure the Blessings of Liberty to ourselves and our Posterity, do ordain and establish this Constitution for the United States of America.[2]

More's *Utopia* (1516) and Skinner's *Walden Two* (1948) use fiction to argue for new and, in those authors' opinions, improved modes of social and political organization.

Huxley's *Brave New World* (1932) and Orwell's *1984* (1949) also use fiction but warn us of their foreboding sense of the future implied in the political choices they see us making.

Most important of all, in that they have influenced the form taken by whole nations, are such efforts as Smith's *Wealth of Nations* (1776), Marx's *Capital* (first volume, 1867), Röpke's

Economics of the Free Society (1937), and Keynes' *General Theory of Employment, Interest, and Money* (1936).

In this sense of influence, I am fascinated by Keynes' concluding paragraphs in the atmosphere of those deep Depression years:

Is the fulfilment of these ideas a visionary hope? Have they insufficient roots in the motives which govern the evolution of political society? Are the interests which they will thwart stronger and more obvious than those which they will serve?

'I do not attempt an answer in this place. It would need a volume of a different character from this one to indicate even in outline the practical measures in which they might be gradually clothed. But if the ideas are correct—an hypothesis on which the author himself must necessarily base what he writes—it would be a mistake, I predict, to dispute their potency over a period of time. At the present moment people are unusually expectant of a more fundamental diagnosis; more particularly ready to receive it; eager to try it out, if it should be even plausible. But apart from this contemporary mood, the ideas of economists and political philosophers, both when they are right and when they are wrong, are more powerful than is commonly understood. Indeed the world is ruled by little else. Practical men, who believe themselves to be quite exempt from any intellectual influences, are usually the slaves of some defunct economist. Madmen in authority, who hear voices in the air, are distilling their frenzy from some academic scribbler of a few years back. I am sure that the power of vested interests is vastly exaggerated compared with the gradual encroachment of ideas. Not, indeed, immediately, but after a certain interval; for in the field of economic and political philosophy there are not many who are influenced by new theories after they are twenty-five or thirty years of age, so that the ideas which civil servants and politicians and even agitators apply to current events are not likely to be the newest. But soon or late, it is ideas, not vested interests, which are dangerous for good or evil.[3]

The countries of the modern industrial world on both sides of the partially opened Iron and Bamboo Curtains are living proof of these observations of Keynes, and their structure, virtues, and problems owe much to the conceptions of these men and their rivals, associates, and followers.

fifteen

If the United States, the advanced nations of Western Europe, and Japan are to be considered prime examples of what we mean by the contemporary industrial society, it takes only relatively limited traveling to conclude that the average citizen in these nations is strikingly better off materially than is the average citizen in the underdeveloped, non-industrialized world. Also, this is so obvious to the citizens of that other world that they are seeking to modify, enlarge, and revolutionize their own societies in a variety of ways so as to generate the capabilities necessary to be included in that industrial world.

Yet, it also is clear that the world we possess, and which they seek, is far from a perfect one. We fight wars; we drink alcohol to excess; we use drugs. The automobiles, which have given us mobility, also jam our city streets, kill and maim in accidents, pollute our air. The sanitation systems we designed to meet another environmental problem, that of the disposition of human wastes, are now all too often contributing to the pollution of our streams and our lakes. The very strengths of our cities as places where men can participate more readily in advantages offered by this industrial world—places where they can find work, schools, shops, theaters, neighborhoods of others like themselves—have attracted so many people into them that their size and congestion threaten to destroy the advantage that brought those same people there. We all know about crime in the streets of those cities. We all know about the almost unbearable tension of the arms race in which we have been and still are engaged with Soviet Russia. We all know about the stresses in our society associated with disengagement from Viet Nam and involvement in the Mid-East war. We all know of the existence of racial prejudice and its consequences. Indeed, a growing minority of citizens are finding the flaws and defects in our system so abhorrent that they challenge in a variety of ways not only some of the values but often the very system itself.

Let me illustrate the nature and the intensity of these criticisms by using the words of a few of the critics themselves. For example, here is Lewis Mumford, writer, social critic, and sometime philosopher in *The Myth of the Machine*:

Even those who see no personal threat from quantification must be prepared to recognize its statistically demonstrable results in the many forms of environmental degradation and ecological unbalance that have resulted from the by-products of our megatechnic economy. The ironic effect of quantification is that many of the most desirable gifts of modern technics disappear when distributed en masse, or when—as with television—they are used too constantly and too automatically. The productivity that could offer a wide margin of choice at every point, with greater respect for individual needs and preferences, becomes instead a system that limits its offerings to those for which a mass demand can be created. So, too, when ten thousand people converge by car on a wild scenic area in a single day 'to get close to nature' the wilderness disappears and megalopolis takes its place.

'In short, megatechnics, so far from having solved the problem of scarcity, has only presented it in a new form even more difficult of solution. Results: a serious deficiency of life, directly stemming from unusable and unendurable abundance. But the scarcity remains: admittedly not of machine-fabricated material goods or of mechanical services, but of anything that suggests the possibility of a richer personal development based upon other values than productivity, speed, power, prestige, pecuniary profit. Neither in the environment as a whole, nor in the individual community or its typical personalities, is there any regard for the necessary conditions favoring balance, growth, and purposeful expression. The defects we have examined lie not in the individual products but in the system itself; it lacks the sensitive responses, the alert evaluations and adaptations, the built-in controls, the nice balance between action and reaction, expressions and inhibitions, that all organic systems display—above all man's own nature.[4]

Here is another viewpoint, that of Charles Reich, professor of law at Yale, as he states it in The Greening of America, a book which, judging by its sales, has had considerable popularity in our colleges and universities:

What is the nature of the social order within which we all live? Why are we so powerless? Why does our state seem impervious to democratic or popular control? Why does it seem to be insane, destroying both self and environment for the sake of principles that remain obscure? Our present social order is so contrary to anything we have learned to expect about a government or a society that its structure is almost beyond compre-

hension. Most of us, including our political leaders and those who write about politics and economics, hold to a picture that is entirely false. Yet children are not entirely deceived, teen-agers understand some aspects of the society very well, and artists, writers and especially moviemakers sometimes come quite close to the truth. The corporate state is an immensely powerful machine, ordered, legalistic, rational, yet utterly out of human control, wholly and perfectly indifferent to any human values. [5]

Or consider this introductory paragraph from B. F. Skinner's *Beyond Freedom and Dignity*:

In trying to solve the terrifying problems that face us in the world today, we naturally turn to the things we do best. We play from strength, and our strength is science and technology. To contain a population explosion we look for better methods of birth control. Threatened by a nuclear holocaust, we build bigger deterrent forces and anti-ballistic-missile systems. We try to stave off world famine with new foods and better ways of growing them. Improved sanitation and medicine will, we hope, control disease, better housing and transportation will solve the problems of the ghettos, and new ways of reducing or disposing of waste will stop the pollution of the environment. We can point to remarkable achievements in all these fields, and it is not surprising that we should try to extend them. But things grow steadily worse and it is disheartening to find that technology itself is increasingly at fault. Sanitation and medicine have made the problems of population more acute, war has acquired a new horror with the invention of nuclear weapons, and the affluent pursuit of happiness is largely responsible for pollution. [6]

Or consider the exchange between Judge Thomsen and a Catholic priest, Father Philip Berrigan, at the trial of the so-called Catonsville Nine, who in 1968 burned the Selective Service draft records of 378 young Catonsville, Maryland, residents:

Your Honor, said Philip Berrigan, I think that we would be less than honest with you if we did not say that, by and large, the attitude of all of us is that we have lost confidence in the institutions of this country, including our own bureaucratic Churches . . . we have no evidence that the institutions of this country, including our own Churches, are able to provide the type of change that justice calls for, not only in this country, but also around the world. [7]

Thus, here are four statements of strong discontent with our society as it is constituted. I have chosen to include so many and at such length to assure that the criticisms are expressed in the articulate phrases of the critics themselves and with sufficient emphasis and detail to establish the atmosphere of confusion and frustration which is endemic to any serious consideration of the organization of our society today. I have purposely omitted statements which have as a basis the dismay, confusion, and frustration generated by the tragic spectacle of a vice president of the United States resigning because of corruption in political office, or the anxieties associated with the implications of Watergate for the White House and the President himself. Rather, I have included those which are critical of the thrust and organization of our whole society, and I could have chosen any of a dozen or a thousand others, many more violent in both words and action.

Self-criticism in our society is nothing new. Indeed, criticism and reaction to criticism have been fundamental to its growth. Nevertheless, it would appear that not since the Civil War has there been so much and such a wide diversity of self-criticism as that in which we have been indulging ourselves this past decade, particularly since some of it has been expressed with so much violence and hate in the form of assassination and riots.

As have most of us, I have listened and reacted to the criticism, sometimes resentfully, sometimes with wonder, sometimes with understanding, often with compassion, often with agreement as to the criticism, but usually not with the cure suggested—if any cure is suggested at all. Each of us undoubtedly has reacted in terms of his knowledge or lack of it, reflected in his background and biases.

Certainly I have had my own pattern. I am an engineer, and for most of my professional life I have been involved in the management of Texas Instruments, an industrial organization emphasizing innovation in its products and services, innovation heavily based on science and technology, emphasizing growth: growth in products and services, growth in diversity, growth around the world. Just about one-third of Texas Instruments products and services are designed and produced directly or indirectly for the Department of Defense, and many of the really significant items of electronic equipment in the Viet Nam War, such as terrain-following radar, real-time infrared for reconnaissance, anti-radar missiles, seismic instrumentation for reconnaissance, laser-guided bombs—all came from our company.

Additionally, as have so many of my general age and professional background, I have been involved in a variety of ways with our Federal Government. I was a Naval Reserve Officer and, as such, spent 3½ years in Washington during World War II. At one time I was vice chairman of the Defense Science Board, a group of advisers to the Department of Defense. I have been a member of a Presidential Commission (the so-called Automation Commission), and the President's Science Advisory Committee, as well as a governor of the new U. S. Postal Service. Presently, I am chairman of the National Council on Educational Research, the policy-making and advisory board of the National Institute of Education.

I have participated, too, in activities concerning my profession—as a member of appropriate professional societies, as a director of the Institute of Radio Engineers and the Institute of Electrical and Electronics Engineers. Indeed, I served as president of the IRE and chairman of its merger committee in the effort that merged it and the American Institute of Electrical and Electronics Engineers to create the IEEE. In addition, I am now a member of the board of trustees and executive committee of each of two universities.

Clearly, I have spent a major proportion of my professional life, and with some considerable degree of responsibility, with the very institutions toward which these criticisms are directed; and to the extent they are justified, I must bear my share of the burden of those criticisms.

Inevitably, I am biased by my experience. I am dismayed by the very real faults in our society and the inadequacies of our overall political system, to say nothing of being deeply shaken personally by the events associated with the resignation of former Vice President Agnew and what at best can be described as the amorality displayed by so many in high places in the whole sequence of events lumped together as the Watergate affair.

But if I have learned anything from my years as engineer and executive, it is that impassioned descriptions or emotional appeals may be useful in calling attention to problems, but they are rarely of very much help in solving them.

It has been my experience that solving problems, or even just improving moderately a negative environmental condition accompanying an otherwise satisfactory solution, is an extraordinarily difficult task when the conflicting needs and views of large numbers of men and women are involved. Progress toward such solutions demands rigorous, non-emotional analysis and usually comes only as the product of very hard work over relatively continuous and long periods of time.

Furthermore, compromises in large variety usually are necessary either to balance the degree to which the conflicting needs and views are met or simply to get anything done at all in the face of widely different perceptions as to the reality or the nature of the problems themselves.

Yet, even though the increasingly frequent observation that we in the United States have been overly concerned about those things contributing to our material welfare and not sufficiently conscious of the overall quality of life in our nation probably is justified, and even though other flaws do exist in our society, some of them serious, I have not seen anywhere else in the world one, in the overall, which to me seems superior. Nor has reasonably extensive reading of history convinced me that past societies were superior for the average man.

These observations do not preclude the possibility that some new combination of structures, attitudes, and mores might indeed provide the more perfect society we are all seeking; but they certainly suggest at least that great caution should be taken in abandoning the patterns and the approaches which have made the United States what it is and that skepticism is a generally justified attitude in the face of most of the doomsday prophecies. Yet, none of us can or wants to deny the flaws that are real, and it is a proper response for each citizen to seek to understand: first, if general observations suggest that the average citizen of the United States is better off materially and enjoys his superiority, too, in life, liberty, and the pursuit of happiness, why that is so; and second, how can we cure the flaws? In addition to providing the material goods and services so abundantly, how can we also improve the quality of life?

In spite of the criticisms of growth, in my opinion the best place to start is by an examination of that growth.

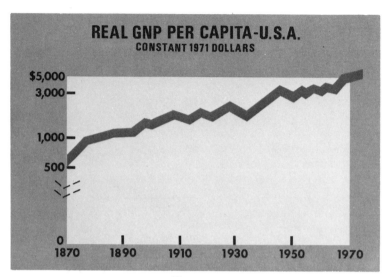

Figure 1[8]

Figure 1 displays what has happened in the United States in the past century, expressed in 1971 dollars. Our gross national product per capita has grown from $661 in 1870 to an estimated $5,674 in 1973, or almost nine times.

Gross national product per capita certainly is not completely equatable with either quality of life or even the quantity of goods and services enjoyed by each citizen. The distribution of wealth affects the latter, and such essential elements as the cleanness of air, clarity of water, or longevity are not assessed by the GNP. Such measures as GNP and GNP per capita often are attacked as having little or no relevance as measures of the quality of life for the average citizen, and economists for whom these are key parameters in their evaluation of the progress of industrial states often are assumed to be ignorant of both the value of these statistics and their shortcomings.

I can think of no better refutation than by quoting from Dr. Simon Kuznets, winner of the Nobel Prize for economic science in 1971 and a pioneer figure in developing in the 1930s for our Department of Commerce these very measures:

> . . . the primary purpose of economic performance and economic growth is to satisfy some positive end-goals of society, not to utilize production potentials merely because they are at

hand. We should therefore ask whether the spread of modern economic growth, as measured by the attainment of some minimum per capita product, reflects the spread of some minimum realized capacity to satisfy the wants and needs consonant with the increased productive power of man.

* * *

'Per capita product thus is only a crude measure of the realized capacity to satisfy an accepted set of end-goals of economic activity, but it is at present one of the few measures available and, in fact, is the only comprehensive one. The measure is therefore generally taken to indicate the extent to which the growing potential of modern technology has been exploited for useful ends, without hopelessly compromising these ends.[9]

But, with full recognition of their limitations, gross national product and gross national product per capita are still the most useful indices we have for measuring and comparing material progress.

Over this century, the United States not only has had a relative gain of eight and one-half times in GNP per capita in spite of a striking reduction in manufacturing working hours (from approximately 62 hours per week in 1870[10] to approximately 40 hours per week in 1973)[11], but over this entire span of time we have enjoyed a much higher level of material prosperity than the other countries of the world.

ESTIMATED 1973 GNP PER CAPITA IN U.S. DOLLAR EQUIVALENTS

United States	$5,980	Czechoslovakia	$1,920
Canada	4,920	Soviet Union	1,800
Sweden	4,910	Portugal	830
Denmark	3,860	Brazil	530
West Germany	3,830	China (Taiwan)	520
France	3,750	Peru	460
Norway	3,720	Tunisia	375
United Kingdom	2,670	Turkey	370
Japan	2,610	Morocco	320
Italy	2,110	Korea (South)	310
Israel	2,080	U.A.R. (Egypt)	230
		India	110

Figure 2[12]

Figure 2 tabulates GNP per capita for countries at all stages of economic development. Although these data are corrected to attempt to reflect actual purchasing power, they are at best an approximation, especially when comparing such widely disparate countries as India and the United States, with India having less than one-fiftieth of our GNP per capita. Nevertheless, the tabulation is useful—first, to show this extreme spread between the developed and undeveloped countries, and second, to point out that among the developed countries the spread is narrowing. Indeed, in many tabulations, using latest exchange rates, both Sweden and Germany show GNPs per capita exceeding that of the United States! At the other end of the scale, there are the somber implications of such countries as India and Egypt, possessing now GNPs per capita which are only a fraction of those attained by the United States a century ago.

Putting aside for the moment debate as to how well GNP per capita identifies with quality of life, there can be little question that any man or woman anywhere in the world would equate the state of his own physical health and the ability to enjoy good health over a long span of life with a high quality of life. A primary measure of physical health is life expectancy.

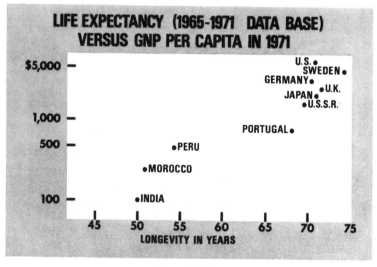

Figure 3[13]

Figure 3 plots life expectancy against GNP per capita, and there is an obvious correlation. India, with the lowest GNP per capita, at $100, also has the lowest life expectancy, at 50 years. Other less developed countries such as Morocco and Peru, with improved GNPs per capita, also have improved life expectancies, although still well behind those of the industrialized nations.

Certainly, there are factors other than material affluence that influence the longevity of a society's citizens. Education and its influence on sanitation is one such factor. Both the physical nature of man himself and the upper bounds of medical knowledge are others.

There are differences, too, introduced in comparing a large, heterogenous society such as that of the United States with a small, relatively homogenous society such as that of Sweden.

Increased wealth also has negative consequences. We use more automobiles, and although our overall accident rate is decreasing, still the increased usage of the automobile means more accident victims every year. Affluence also makes it too easy for too many of us to be overweight with a consequent marked increase in heart and other disease.

But even considering all of these other factors, it still is clear

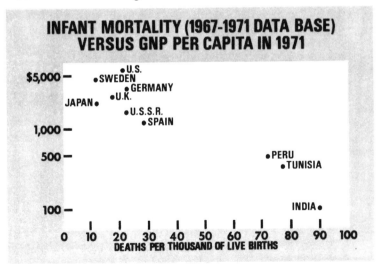

Figure 4[14]

that the overall wealth of a society affects the state of its technology, the education of its citizens, its ability to provide sanitation, combat infectious disease and ensure adequate diets, and in general, add to life expectancy.

Another basic indicator of the state of general health is infant mortality. (See Figure 4.)

Once again the correlation, while not perfect, is very strong. India couples the highest infant mortality rate with the lowest GNP per capita. The high-standard-of-living countries also show the lowest infant mortality rates—in general about one-fifth to one-seventh of India's.

Just because there are differences among countries and races that complicate comparisons, examining what has happened here in the United States over the past 40 years may be more meaningful.

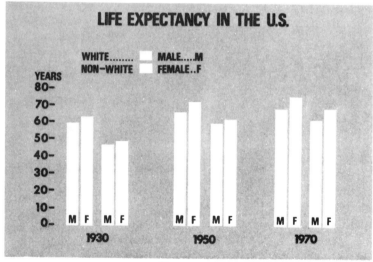

Figure 5[15]

Figure 5 portrays the life expectancies in 1930, 1950, and 1970 for whites and non-whites. The female superiority over the male is marked for both whites and non-whites. But all have made sharp gains, ranging from more than eight years (or a 14% increase) for the white male to more than 19 years (or a 40% increase) for the non-white female over the 40-year span. In fact, in 1970, the life expectancy of the non-white female had improved sufficiently to exceed that of the white male.

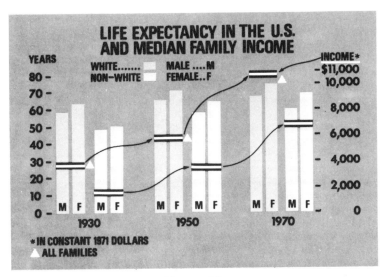

Figure 6[16]

In Figure 6 the median income per family in 1971 dollars has been superimposed over the bar chart of life expectancies. Median income of all families has grown approximately 2.8 times, to $10,289 in 1970, up from an estimated $3700 in 1930.

The median income of white families has been slightly higher in each year and has grown in approximately the same proportion. Although the median income of non-white families has been and remains lower than that for white families, its growth has been significantly higher. The level exceeded $6800 in 1970, about the same as that of whites, as recently as 1955, more than doubling since 1950 and almost quadrupling the comparable estimated level of $1770 in 1930. Again, there certainly appears to be a strong correlation between improving income and improving life expectancy.

Another measure of quality of life for the average citizen is his education. What distinguishes a civilized man from a barbarian is surely his ability to transmit knowledge — knowledge in the sciences, in the arts, in the society's own traditions from one generation to the next. Although literacy is not a completely adequate measure of education, still it must be accepted as a reasonable approximation of the degree to which a broad base of primary education is available.

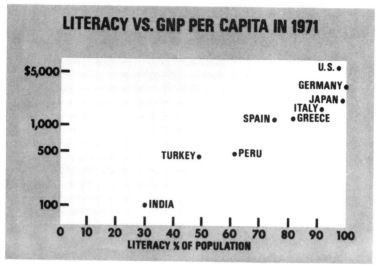

Figure 7[17]

Figure 7 plots literacy as a percent of population against GNP per capita. Again, it is clear that there is a considerable correlation between material affluence of the society and both its ability to provide a broad base of formal education for its people and the inclination and likelihood of those citizens to take advantage of that availability. Note India at the bottom with only a 29% literacy rate against a GNP per capita of $100 and the United States and other industrialized countries with literacy rates approaching 100% against very high GNPs per capita.

Once again, a comparison restricted to the United States may be even more meaningful. (See Figure 8.)

In the 42 years between 1930 and 1972, the average American over 25 years old, who in the earlier span of time had an eighth-grade education, became a high school graduate. The increasing level of education accompanied and undoubtedly contributed to the increasing standard of living.

Sometimes, when one is barraged by the outpouring of criticisms of the industrialized societies, one would have to conclude that the deterioration of the overall quality of life accompanying the increasing affluence in those same societies has been overwhelming and that such a deterioration is an inevitable accompaniment of the increase in material welfare. There are very real problems, but the overall gains in quality of life have been enormous. With all the emphasis on the problems of our cities, one would certainly be led to believe that there has been decay in all aspects of their quality of life as they have grown. Yet, until recently, death rates were much higher in the cities than they were in the rural areas. In London, the death rate from 1701 through 1750 was 49 per thousand against a rate of 33 per thousand for all of England and Wales.[19] As late as 1830, death rates in Boston, New York, and Philadelphia were twice as high for age groups from 5 to 15 and 19 to 79 as they were in 44 rural townships in New England.[20] The situation today is just the opposite, and there is no developed country that has a death rate unfavorable to the urban population.[21]

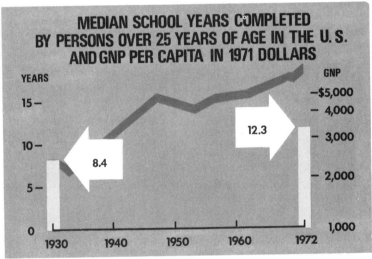

Figure 8[18]

As another instance, I remember vividly being in London on a winter day in the 1950s when one of London's notorious black fogs enveloped the city. I was literally unable to see my hand when I held my arm outstretched before me. My destination was only a block or two away from the Savoy Hotel on the Strand. Yet, it took me more than 20 minutes to make that walk, feeling my way along the building walls and avoiding other pedestrians who were doing the same thing.

All of that has been changed today. There was an annual average of 95 hours of thick and dense fog in the years 1947 to

twenty-nine

1954. This decreased to just 17 hours average annually for the years 1963 through 1969. The average winter sunshine in London these days is 50% above the long-term average for the years before 1960. (See Figure 9.)

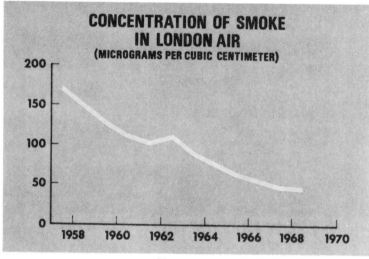

Figure 9[22]

Specifically, in the 14 years after the passage of the Clean Air Act in 1956, the average smoke concentration fell from above 150 micrograms per cubic meter to below 50.[23]

But who knows better than Pittsburgh citizens what a determined effort uniting all segments of the community and applying available technology can accomplish? Figure 10 is a photograph showing what Pittsburgh looked like when it was known as the "Smoky City."

The gains since the Allegheny County Smoke Control Ordinance was passed in 1949 are astonishing. Available reports indicate that, in 1940, downtown visibility was below three-quarters of a mile for more than 1000 hours. This had been dropped to fewer than 75 hours per year by 1960, and, in 1970, visibility was more than 10 miles most of the time.[25]

On a broader scale, any organized national effort to improve environmental quality in the United States is very recent indeed.

Yet, even in these few years, in the problem area where perhaps we are best able to describe the problem (that of air pollution in the cities) we appear to be making progress. Con-

Figure 10[24]

sider the Mitre Air Quality Index, which combines indices for individual pollutants for each site relating the measured levels of pollution to the air quality standards of the Environmental Protection Agency. (Figure 11.)

There was a 14% improvement in the 83 sites sampled between 1968 and 1969 and another 10% improvement between 1969 and 1970.

Perhaps even more significant is the Extreme Value Index, which measures the extent of very high level pollution for very short periods of time. The pollution conditions measured (Figure 12) are those most directly related to personal comfort and well being.

The improvement here was even more striking, nearly 32% between 1968 and 1969, and an additional nearly 17% between 1969 and 1970.

I must hasten to add that, in its report, the Council on Environmental Quality emphasizes the tentative nature of these indices and how many questions still must be answered as to the effectiveness with which they are measuring the significant fac-

MAQI* VALUES, 1968-1970
BY POPULATION CLASS

POPULATION CLASS OF COMMUNITY	1968 MAQI	1969 MAQI	PERCENT IMPROVEMENT 1968-69	1970 MAQI	PERCENT IMPROVEMENT 1969-70
NATIONAL SAMPLE 82 SITES	3.77	3.24	14.1	2.91	10.2

*MITRE AIR QUALITY INDEX

Figure 11[26]

tors. Indeed, this 1972 report includes the following disclaimer: "Neither this year nor next will we be able to provide a general statement about whether environmental quality has improved or deteriorated. The environment encompasses too many fac-

EVI* VALUES, 1968-1970
BY POPULATION CLASS

POPULATION CLASS OF COMMUNITY	1968 EVI	1969 EVI	PERCENT IMPROVEMENT 1968-69	1970 EVI	PERCENT IMPROVEMENT 1969-70
NATIONAL SAMPLE 82 SITES	10.32	7.03	31.9	5.84	16.9

*EXTREME VALUE INDEX

Figure 12[27]

tors to be so easily characterized . . . ."[28]

It has to be significant, however, that we appear to be making progress in just the area where we are beginning to be able to identify and define the problem, which is not surprising to anyone who has struggled to solve any kind of complex problem.

I suggest that we often yearn for "good old days" that never were. One such demonstration is Don Berkebile's amusing yet sobering article in a recent edition of the Smithsonian Magazine. (See Figure 13.)

Figure 13[29]

Here are a few quotes:

Deep within the breast of many crusaders for a better environment lies the thought that the automobile is the true root of all our evils, that inventing it in the first place was the devil's own work and that becoming enslaved by it, as we surely are, has opened the way to our destruction. Back to the horse and buggy, we sometimes murmur as we choke for oxygen on a jammed city street. Back to the days of breathable air, when a person could cross a street safely and with dignity and could park his horse and carriage within mere feet of his destination.

'Yet a cool appraisal of horse-and-buggy days casts doubts on the ecological value of the old gray mare. Consider pollution:

thirty-three

London, in 1875, had to get 1,000 tons of horse manure off its streets—daily. American cities with a population of 12,000 horses (New York had more than ten times as many in the turn of the century) had to remove, daily, a 130-ton hill of horse manure.

* * *

'While waiting to be cleaned up from the streets, manure bred billions of flies which carried at least 30 diseases, some of them quite serious. . . .

* * *

'Dead horses, of course, presented another problem. Wherever there was a huge horse population, there were sure to be huge numbers of corpses that had to be dragged away, at great effort and expense. In New York and Chicago, around 1900, as many as 15,000 horses a year had to be hauled off. . . .

'Carriages, with their high center of gravity and uncertain means of locomotion, were accident-prone. Late in the 19th century, Paris could count 700 deaths and 5,000 injuries a year from carriage accidents. . . .

'As for parking a horse and carriage, remember that the smallest horse-drawn vehicle (four-wheeled) filled a space of roughly 6 by 16 feet, including the horse. Many such buggies could only carry two people. Compare that to today's automobile—6 by 17 or 18 feet—carrying six people. . . .

'The cost and problems of horse transport in cities during the early years of this century were so great that many of our forebears looked upon the automobile as their only salvation from *urban ills and limitations. . . . .*[30]

* * *

But to switch away from the cities and their problems to the countryside, I had gained the impression that we in the United States were depleting our forest resources at a high rate. Imagine my surprise, then, to learn that in spite of the growth in population and intensive cultivation of much of the country, we have 75% as much forest land today as we did when Columbus landed. An American Forest Institute report lists our total area of forested land as 758 million acres, and one-third of that is set aside in parks, wilderness areas, and watersheds, or is not suitable for commercial timber. Another startling fact to me is how heavily forested our Eastern states are. In fact, our biggest gains in forest lands have come in New England, where the gradual abandonment of agriculture on unproductive lands has resulted in reforesting of fields and other clearings. For example, 90% of Maine's acres and 89% of Vermont's are forested; Connecticut has 70% of its acreage in forests; Rhode Island, 65%; and heavily populated New York State has 57%.[31]

Or to turn to a completely different demonstration of an improving quality of life, Figure 14 breaks out some of the information about education in the United States contained within that satisfying statistic previously displayed (Figure 8), which made the point that the average American above 25 years old was a grade school graduate in 1930 and by 1972 had become a high school graduate.

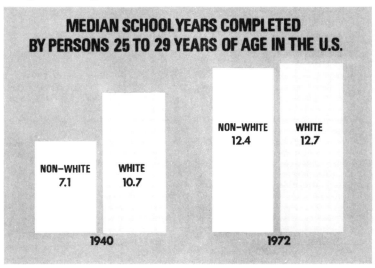

Figure 14[32]

The shift is much more striking when we see the gains made by non-whites. Ten years later, in 1940, non-whites between 25 and 29 years of age still had not quite come to the overall eighth grade level, having completed just 7.1 years of school against 10.7 for the whites. In 1972, for this same age group, the gap had practically been closed with 12.4 years average for non-whites against 12.7 years for whites!

Or consider the number of college degrees granted per year (Figure 15)—from fewer than 10,000 in 1870 to more than a million in 1970.

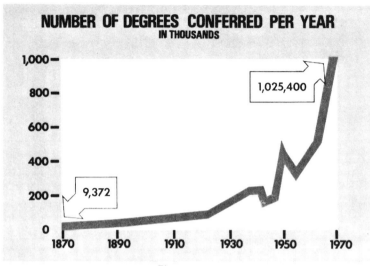

NUMBER OF DEGREES CONFERRED PER YEAR
IN THOUSANDS

1,025,400

9,372

Figure 15[33]

All of these are just a few of the overwhelming number of examples of improvements in our quality of life accompanying increased national and personal material wealth.

Certainly, there are problems and some of them are not going to be readily solved. Often, the legislative and technological solutions we apply are not the best, and the costs associated with their application seem larger than the gains made. Nevertheless, it is extraordinarily difficult not to conclude, first, that such fundamental aspects requisite to a life of quality as good health and education are found in the high, not the low, productivity societies, and second, that we have been and are making progress in continuing to improve the quality of our lives, including our environment.

Dr. Daniel Boorstin, presently director of the National Museum of History and Technology, has expressed extraordinarily well in one of his recent books the relationship between the quality of life for every citizen and a high standard of living. He says:

The growth of Consumption Communities has signaled a transformation of the attitude to all material goods. By contrast with the rigid, Old World notion of wealth, the New World idea of standard of living has had certain obvious characteristics.

'Standard of living is public. But it is possible to be 'wealthy' in secret. A man can hide his treasure in a vault, in his garden, in a mattress. . . . It is not possible to have a high standard of living in secret.

'The word 'standard' (which comes into English from the Old French estandard, 'banner') means a symbol that is displayed for all to see. Its very function is to be seen, to inform as much as to affirm. A standard of living then, is a publicly seen and known measure of how people do live, and of how they should live. . . .

* * *

'Because a high standard of living is a public fact, it becomes a public benefit. You can become rich without my becoming richer. But it is hard for you to have a high standard of living without incidentally raising mine. . . . . If, in addition to your material goods, your standard of living includes your freedom from threat of crime or disease, your education, the education of your children, the air you breathe, the water you drink, the roads you drive on, the public transportation system you use, your peace of mind—then does it not inevitably include my opportunities and the opportunities of my children for education (in the institutions you support), the air I breathe, the water I drink, the roads I drive on, the public transportation I use, and my freedom from threat of crime or disease?[34]

Dr. Boorstin's statements help us to appreciate that such measures as gross national product and total disposable income (which we use to describe our standard of living because they are what is available) obviously fail to express what really is meant by a standard of living.

It was work sponsored by the National Bureau of Economic Research in the 1920s that led directly to our national income estimates and the entire collection of immensely useful data encompassed by the term, "gross national product," that we take so for granted today, and it is certainly appropriate that NBER now is sponsoring studies seeking an expanded ability to comprehend and measure more adequately our standard of living.

In their 1972 paper, "Is Growth Obsolete?" prepared as a part of an NBER study, William Nordhaus and James Tobin state:

thirty-seven

A major question raised by critics of economic growth is whether we have been growing at all in any meaningful sense. Gross national product statistics cannot give the answers, for GNP is not a measure of economic welfare. . . . .

'An obvious shortcoming of GNP is that it is an index of production, not consumption. The goal of economic activity, after all, is consumption. Although this is the central premise of economics, the profession has been slow to develop, either conceptually or statistically, a measure of economic performance oriented to consumption, broadly defined and carefully calculated. We have constructed a primitive and experimental 'measure of economic welfare (MEW),' in which we attempt to allow for the more obvious discrepancies between GNP and economic welfare. . . . [35]

Nordhaus and Tobin adjust GNP by reclassifying GNP expenditures into consumption, investment, and intermediate products; by imputing dollar values for the services of consumer capital; for leisure and for the product of household work; and finally by subtracting some of the disamenities of urbanization.

GNP AND MEASURES OF ECONOMIC WELFARE
BILLIONS OF DOLLARS, 1958 PRICES

	1935	1947	1965
GROSS NATIONAL PRODUCT.......	169.5	309.9	617.8
Capital consumption, NIPA.........	−20.0	−18.3	−54.7
Net National Product, NIPA.........	149.5	291.6	563.1
Regrettables and intermediates:			
Government...............	−7.4	−20.8	−63.2
Private..................	−9.2	−10.9	−30.9
Imputations for:			
Leisure.................	401.3	466.9	626.9
Non-market activity..........	109.2	159.6	295.4
Disamenities...............	−14.1	−19.1	−34.6
Public and private capital......	24.2	36.7	78.9
Additional capital consumption......	−33.4	−50.8	−92.7
GROWTH REQUIREMENT..........	−46.7	+5.4	−101.8
SUSTAINABLE MEW.............	573.4	858.7	1241.1

Figure 16[36]

Figure 16 displays the results of these adjustments to GNP for the years 1935, 1947, and 1965, using 1958 prices. It recognizes that the depreciation of capital stocks is a cost of production and must be subtracted to arrive at net national product. NNP grew from $149.5 billion in 1935 to $563 billion in 1965.

Because NNP still is a measure of production, rather than of consumption or welfare, it includes inputs that are necessary, but do not in themselves provide useful products. For example, the cost of commuting to work—or such items as police, sanitation and road maintenance services, and national defense —are all classified as "ancillary, regrettable" activity. Nordhaus and Tobin estimate the total of this kind of overhead expense rose from just under 10% of GNP in 1935 to 15% in 1965.

Again, because net national product measures only market activity, leisure and such nonmarket productive activity as work in the home or school are not included. Yet, obviously, welfare rises even though the effect on net national product is negative when we choose voluntarily, as we have been doing for decades, to work fewer hours per week, fewer weeks per year, and fewer years per lifetime.

In this summary, leisure, growing from $401 billion in 1935 to $627 billion in 1965, is deflated by the change in wage rate. On the other hand, the assumption is made that nonmarket productive activity is augmented by technological progress to the same degree as is marketed labor; hence, the value of the non-market activity is deflated by the usual consumption deflator. You will note that non-market activity nearly triples over the 30 years to $295 billion.

Many of the negatives, so emotionally described in such criticisms as those with which I began this discussion, are by-products of urbanization and congestion. The gains in net national product have been made possible to a major extent by the vast migration of workers from the farm to urban industry. This summary assumes that some proportion of the higher earnings accompanying that migration is pay for these disamenities of urban life and work. There is a persistent association of higher wages with higher population densities, and Nordhaus and Tobin have applied cross-sectional regressions to identify the income differentials necessary to attract and keep people where the population densities are high. The estimate so obtained: $14 billion in 1935 and nearly $35 billion in 1965, is approximately 5% of GNP.

Although net national product already deducts depreciation of capital stocks, it fails to treat other durable goods, such as consumer durables, as capital; and this, too, is adjusted in arriving at MEW. Finally, education and health expenditures, both public and private, are treated as capital investments. To the extent, then, that education and medical care are direct sources of consumer satisfaction, rather than capital, MEW is understated.

For a stationary population, the net national product, since depreciation has been subtracted, measures the consumption level the economy could sustain indefinitely. However, it is not aggregate consumption but per capita consumption that determines the welfare of the individual. Because our population is growing, the capital stock must grow at the same rate as the population in the labor force. This is a real requirement and, for a growing population, a true cost of staying at the same level of welfare.

In a society such as ours, however, with a high level of technological progress, the sustainable consumption per capita grows steadily, and the growth requirement in Figure 16: $46.7 billion in 1935 and nearly $102 billion in 1965, takes the requirement for technological progress, as well as growth in population and labor force, into consideration. This is a very important correction, since it is nearly one-sixth of the GNP in 1965.

Correcting for the growth requirement finally gives us a sustainable measure of economic welfare: $573 billion in 1935, $858 billion in 1947, and $1241 billion in 1965.

Actual MEW may be something quite different from this sustainable MEW. If the growth requirements are met, per capita consumption can grow at the trend rate of increase in labor productivity. If the actual MEW is less than the sustainable MEW, then as a society, we are providing even better for our children as prospective consumers than we are for ourselves. If actual MEW exceeds sustainable MEW, however, our current consumption is, in effect, consuming some of the fruits of possible future progress.

In Figure 17, Nordhaus and Tobin, using the same approach, have arrived at total consumption, that is, the actual MEW,

MEASURES OF ECONOMIC WELFARE (MEW)
ACTUAL AND SUSTAINABLE
BILLIONS OF DOLLARS, 1958 PRICES

	1935	1947	1965
Personal consumption.	125.5	206.3	397.7
Private instrumental expenditures, durable goods, and other household investment.	−27.0	−47.5	−121.9
Consumer capital imputation.	17.8	26.7	62.3
Imputation for leisure.	401.3	466.9	626.9
Imputation for non-market activities. . .	109.2	159.6	295.4
Disamenity correction.	−14.1	−19.1	−34.6
Government consumption.	0.3	0.5	1.2
Government capital imputation.	6.4	10.0	16.6
Total Consumption =ACTUAL MEW.	**619.4**	**803.4**	**1243.6**
MEW net investment.	−46.0	55.3	−2.5
SUSTAINABLE MEW. . . .	**573.4**	**858.7**	**1241.1**

Figure 17[37]

growing from $619 billion in 1935 to $1244 billion in 1965. Comparing this actual MEW with the sustainable MEW (see Figure 16) one can identify the net investment in MEW in the future or consumption at a rate which is at the cost of future progress. Thus, in 1935, MEW net investment was a negative $46 billion; in 1947, a positive $55 billion; and in 1965, sustainable and actual MEW were very close together at $2.5 billion.

The authors have stressed that they consider this a very primitive attempt to measure economic welfare and have attempted to assess the reliability of the estimates making up their MEW. While the reliability of our gross national product figures have never been studied to determine what kind of statistical confidence factor exists, estimates are that any errors are probably less than 3%. In their judgment, errors in the various elements of MEW vary from perhaps twice that which may be present in gross national product (where they have corrected for instrumental expenditures) to errors that probably are 10 times as great as those in the GNP in their imputations for leisure and non-market activity and disamenities.

In welfare, what really counts is the per capita, not the aggregate level.

Figure 18 compares actual and sustainable MEW per capita

MEASURES OF ECONOMIC WELFARE (MEW)
ACTUAL AND SUSTAINABLE

	1935	1947	1965
ACTUAL MEW PER CAPITA			
IN DOLLARS.............	4866	5552	6391
1929 = 100............	108.0	123.2	141.8
SUSTAINABLE MEW PER CAPITA			
IN DOLLARS.............	4504	5934	6378
1929 = 100............	100.9	133.0	142.9
NET NATIONAL PRODUCT			
DOLLARS PER CAPITA.......	1205	2038	2897
1929 = 100............	78.0	131.9	187.5

Figure 18[38]

with per capita net national product. Again in 1965, the sustainable per capita MEW and the actual per capita MEW are nearly the same at $6,391 and $6,378, respectively. Both are well above the per capita NNP of $2,897. Clearly, then, MEW is something quite different from most conventional output measures, and consumption items omitted from the GNP are of real significance.

Finally, MEW has been growing at 1.1% per capita per year as compared to 1.7% for NNP. Although MEW has been growing more slowly than NNP, it has been growing, and the authors conclude: *"The progress indicated by conventional national accounts is not just a myth that evaporates when a welfare-oriented measure is substituted."*[39]

But having come to that conclusion, there is another group of critics who say: "Of course, we have attained a higher standard of living and an improved quality of life through industrialization and the higher productivity society it allows. But all of this is temporary. The world's resources are becoming exhausted and will limit our ability to sustain it."

For example, there is J. F. Forrester's *World Dynamics*, which he describes as a preliminary effort to construct a suitable methodology to improve comprehension and deal at least in a broad way with the changes accruing in the world.

Here are a few paragraphs from its introduction:

1. *Industrialization may be a more fundamental disturbing force in world ecology than is population. In fact, the population explosion is perhaps best viewed as a result of technology and industrialization.*

* * *

3. *We may now be living in a 'golden age' when, in spite of a widely acknowledged feeling of malaise, the quality of life is, on the average, higher than ever before in history and higher now than the future offers.*

4. *Exhortations and programs directed at population control may be inherently self-defeating. If population control begins to result, as hoped, in higher per capita food supply and material standard of living, these very improvements may relax the pressures and generate forces to trigger a resurgence of population growth.*

* * *

7. *A society with a high level of industrialization may be nonsustainable. It may be self-extinguishing if it exhausts the natural resources on which it depends. Or, if unending substitution for declining natural resources were possible, a new international strife over pollution and environmental rights might pull the average world-wide standard of living back to the level of a century ago.*

8. *From the long view of a hundred years hence, the present efforts of underdeveloped countries to industrialize may be unwise. They may now be closer to an ultimate equilibrium with the environment than are the industrialized nations.*[40]

All of these statements seem to flow from the mathematical models constructed to portray the world's condition some decades hence, and here are some of the conclusions drawn:

The world is running away from its long-term threats by trying to relieve social pressures as they arise. But, if we persist in treating only the symptoms and not the causes, the result will be to increase the magnitude of the ultimate threat and reduce our capability to respond when we no longer have more space and resources to invade.

'What does this mean? Instead of automatically attempting to

forty-three

cope with population growth, national and international efforts to relieve the pressures of excess growth must be reexamined. Many such humanitarian impulses seem to be making matters worse in the long run. Rising pressures are necessary to hasten the day when population is stabilized. Pressures can be increased by reducing food production, reducing health services, and reducing industrialization. Such reductions seem to have only slight effect on quality of life in the long run. The principal effect will be in squeezing down and stopping the runaway growth.

'The limitation of capital investment may be even harder to achieve than a limit on population. There is less recognition of industrialization as a threat to the environment than there is of population. The temporarily higher standard of living that often comes from industrialization is now sought by most cultures. Damage to the environment is being caused by technological processes, but in spite of this, hope is placed instead on population control. Even if population control were achieved, rising industrialization would lead into trouble. Conversely, unless industrialization is limited directly, population control probably cannot be achieved. [41]

This is extraordinary counsel. It says, in effect, that all of man's efforts and struggles to increase his food supply, to cure disease, to assure good health—generally to enjoy the good life—may be counterproductive and, in the end, assure just the opposite.

Because the conclusions drawn are based on mathematical models manipulated by a computer, it may appear that they avoid the emotional bias characteristic of some of the criticisms I quoted earlier in this discussion. However, the conclusions are not likely to be better than the models, which generally assume that natural resources or the earth's ability to absorb waste are finite, and that a world economy based on industrialization will consume the natural resources at an ever-increasing rate and will saturate the environment with its waste products at the same time. There are no mechanisms in the model that decrease the consumption of any one resource or, for that matter, the creation of waste products gradually.

Economist Robert Solow has said it very well:

The most glaring defect of the Forrester-Meadows models is the absence of any sort of functioning price system. The price system is, after all, the main device evolved by capitalist (and, to an increasing extent, even socialist) economies for registering and reacting to relative scarcity. I have already mentioned one way that a price system might radically alter the behavior predicted by the models—by inducing more active search for resource-saving innovations as resource costs bulk larger in total costs and appear to be increasing. There are other, more pedestrian, modes of operation of the market mechanism. Higher and rising prices of exhaustible resources can be expected to lead to the substitution of other, more plentiful, and, therefore, cheaper materials. To the extent that it is impossible to design around or substitute away from expensive natural resources, the prices of commodities containing a lot of them will rise relative to the prices of other goods and services.

'Consumers will be driven to buy fewer resource-intensive goods and more of other things. All these effects work automatically to increase the productivity of natural resources, i.e., to reduce resource requirements per unit of GNP, and steadily, not once-and-for-all — indeed, one might say 'exponentially' as first approximation.

'To forestall misunderstanding, let me say that this is not an argument for laissez-faire. Many markets are 'imperfect'; they contain substantial elements of monopoly; information is of different quality and spread unevenly among participants; and private interests, to which prices respond, may be in conflict with the public interest. Since the proper response depends to some extent on what is expected to happen and not on what has already happened, uncoordinated activity may lag too far behind events, and even become perverse. Public agencies can try to shorten the lag by providing the best available information about future supplies and demands.

'But I don't see how one can have the slightest confidence in the predictions of models that seem to make no room for everyday market forces.

'As a matter of fact, the relative prices of natural resources and resource products have shown no tendency to rise over the past half-century. This means there have been offsets to any pro-

gressive impoverishment of deposits – improvements in extraction technology, savings in use, the availability of cheaper substitutes. The situation could, of course, change. If the expert participants in the market now believed that resource prices would be sharply higher at some time in the future, prices would already be rising.[42]

* * *

'I have heard it said that, even if the Doomsday Models are all balderdash, the publicity surrounding them has served an important social purpose in drawing the world's attention to the possibility that Spaceship Earth is about to abort. Maybe so. It seems to me more likely, however, that the net effect will be a minus. A sound analysis of the dynamics of population growth, resource use, and environmental pollution might provide the basis for public policy. It might, that is, suggest a list of things to do that can actually be done, and say what will happen if they are done or if they aren't.

'My impression is that the Doomsday Models divert attention from remedial public policy by permitting everyone to blame 'the predicament of mankind.' Who could pay attention to a humdrum affair like legislation to tax sulfur emissions when the date of the Apocalypse has just been announced by a computer?[43]

The prevailing standard economic model of growth assumes that production depends only on labor and reproducible capital. The third factor of input, land and resources generally has been dropped. Essentially, the judgment is made that reproducible capital is a near-perfect substitute for land and other exhaustible resources and that innovation will overcome resource scarcity or exhaustion.

Obviously, in the minds of those environmentalists who hold that no substitutes are available for natural resources, these general economic assumptions about technology and innovation are far too optimistic. The assumption made as to substitutability between capital and labor and natural resources is then the key question. Nordhaus and Tobin, in the paper on "Economic Growth" to which I referred earlier, also took a preliminary look at what the evidence suggests. Here are their conclusions:

First we ran several simulations of the process of economic growth in order to see which assumptions about substitution and technology fit the 'stylized' facts. The important facts are: growing income per capita and growing capital per capita; relatively declining inputs and income shares of natural resources; and a slowly declining capital-output ratio. Among the various forms of production function considered, the following assumptions come closest to reproducing these stylized facts: (a) Either the elasticity of substitution between natural resources and other factors is high—significantly greater than unity—or resource-augmenting technological change has proceeded faster than overall productivity; (b) the elasticity of substitution between labor and capital is close to unity.

'After these simulations were run, it appeared possible to estimate directly the parameters of the preferred form of production function. Econometric estimates confirm proposition (a) and seem to support the alternative of high elasticity of substitution between resources and the neoclassical factors.

'Of course it is always possible that the future will be discontinuously different from the past. But if our estimates are accepted, then continuation of substitution during the next fifty years, during which many environmentalists foresee the end to growth, will result in a small increase—perhaps about 0.1 per cent per annum—in the growth of per capita income.[44]

Some of the zero-growth conservationists no doubt would refuse to accept this analysis on the grounds that our mixed economy is naturally wasteful of resources and that the market simply doesn't recognize a wide variety of values adequately in its pricing system. The fallacy of that position can best be seen by considering the two basic cases. First, there are the usual and identifiable resources for which buyers pay market values and users market rentals. Second, there are the public goods, like clean air and clean water, whose use appears free to individual producers and consumers but is costly in aggregate to society. Nordhaus and Tobin treat this as follows:

If the past is any guide for the future, there seems to be little reason to worry about the exhaustion of resources which the market already treats as economic goods. We have already commented on the irony that both growth men and antigrowth men invoke the interests of future generations. The issue between them is not whether and how much provision must be made for future generations, but in what form it should be made. The growth man emphasizes reproducible capital and education. The conservationist emphasizes exhaustible resources—minerals in the ground, open space, virgin land. The economist's initial presumption is that the market will decide in what forms to transmit wealth by the requirement that all kinds of wealth bear a comparable rate of return. Now stocks of natural resources—for example, mineral deposits—are essentially sterile. Their return to their owners is the increase in their prices relative to prices of other goods. In a properly functioning market economy, resources will be exploited at such a pace that their rate of relative price appreciation is competitive with rates of return on other kinds of capital. Many conservationists have noted such price appreciation with horror, but if the prices of these resources accurately reflect the scarcities of the future, they must rise in order to prevent too rapid exploitation. Natural resources should grow in relative scarcity—otherwise they are an inefficient way for society to hold and transmit wealth compared to productive physical and human capital. Price appreciation protects resources from premature exploitation.

'How would an excessive rate of exploitation show up? We would see rates of relative price increase that are above the general real rate of return on wealth. This would indicate that society had in the past used precious resources too profligately, relative to the tastes and technologies later revealed. The scattered evidence we have indicates little excessive price rise. For some resources, indeed, prices seem to have risen more slowly than efficient use would indicate ex post.

'If this reasoning is correct, the nightmare of a day of reckoning and economic collapse when, for example, all fossil fuels are forever gone seems to be based on failure to recognize the existing and future possibilities of substitute materials and processes. As the day of reckoning approaches, fuel prices will provide—as they do not now—strong incentives for such substitutions, as well as for the conservation of remaining supplies.[45]

I suppose that in these days of energy shortages, the self-defeating nature of price controls, which actually encourage the overuse of scarce resources by undervaluing them, is obvious.

The authors agree that such abuses of public natural resources as air pollution, water pollution, noise pollution, and visual disamenities present a much more serious problem.

You will recall that their measure of economic welfare (MEW) estimates included a disamenity charge of about 5% of total consumption for urban disamenities to compensate for the overestimation of welfare produced in the national product series. They then go on to say:

There are other serious consequences of treating as free things which are not really free. This practice gives the wrong signals for the directions of economic growth. The producers of automobiles and of electricity should be given incentives to develop and to utilize 'cleaner' technologies. The consumers of automobiles and electricity should pay in higher prices for the pollution they cause, or for the higher costs of low-pollution processes. If recognition of these costs causes consumers to shift their purchases to other goods and services, that is only efficient . . .

'The mistake of the antigrowth men is to blame economic growth per se for the misdirection of economic growth. The misdirection is due to a defect of the pricing system—a serious but by no means irreparable defect and one which would in any case be present in a stationary economy . . . The proper remedy is to correct the price system so as to discourage these technologies. Zero economic growth is a blunt instrument for cleaner air, prodigiously expensive and probably ineffectual.[46]

Common to many of the doomsday prophesies is an apocalyptic vision of the earth crushed by teeming billions of fecund humans, reproducing until our supplies are exhausted.

Further, they contrast this catastrophic consequence of increasing population with the common assumption that growth in population contributes to the improvement of a society's material welfare or, indeed, that it depends upon it. To some extent, these latter misconceptions arise because the neoclassical economic model does assume that population and labor force grow exogenously, just like compound interest.

Since it is clear that there is some adaptation both of human fertility and mortality to economic circumstances, the model is due for revision. A market system is certainly not set up to reflect the costs to society of additional children, which may well exceed the costs to the parents. Internalizing the full social costs of reproduction will be far more difficult than the revisions to the market system that might be necessary to internalize the social costs of pollution.

Yet, as Nordhaus and Tobin point out, during the last 10 years, the fertility of the U. S. population has declined dramatically, and, if the trend continues, fertility would soon be diminished to a level consistent with zero population growth.

U.S. POPULATION CHARACTERISTICS IN EQUILIBRIUM

	INTRINSIC GROWTH RATE (% PER YEAR)	NET REPRODUCTION RATE	MEDIAN AGE
1960 FERTILITY — MORTALITY	2.1362	1.750	21-22
1967 FERTILITY — MORTALITY	0.7370	1.221	28
1972 FERTILITY — MORTALITY (EST.)—	0.2000	0.960	38
HYPOTHETICAL ZPG	0.0000	1.000	32

Figure 19[47]

Figure 19 compares the intrinsic or equilibrium growth rates for the net age specific reproduction rates of 1960, 1967, 1972, and for zero population growth. Note that between 1960 and 1967 the growth rate dropped from more than 2% per year to below 0.75% per year.

Further, the estimate is that the intrinsic growth rate dropped to below zero in 1972, producing a net reproduction rate of .960, less than that required to arrive at zero population growth. Assuming, however, that this might represent a slight overshoot and that the actual net reproduction rate will stabilize at one, the population still would go on growing slowly for another half century, perhaps, while the bulge in the age distribution, a

POPULATION CHARACTERISTICS
OF OTHER INDUSTRIALIZED NATIONS

COUNTRY	TIME PERIOD	INTRINSIC ANNUAL GROWTH RATE	NET REPRODUCTION RATE
CANADA	1955-59	2.17	1.82
	1971	0.09	1.03
JAPAN	1947-49	1.85	1.73
	1960-64	−0.33	0.91
	1967	0.16	1.05
GERMANY	1955-59	0.12	1.04
	1965-69	0.44	1.13
FRANCE	1956-60	0.87	1.27
	1969	0.66	1.20
SWEDEN	1955-59	0.20	1.06
	1970	−0.30	0.92
BELGIUM	1960-64	0.75	1.23
	1969	0.20	1.06
NETHERLANDS	1955-59	1.26	1.46
	1968	0.91	1.29
UNITED KINGDOM	1955	0.45	1.13
(ENGLAND & WALES)	1960-69	1.01	1.31
	1971	0.46	1.13

Figure 20[48]

product of the more fertile years, dissipated. The population of the United States then would level off between 250 and 300 million, hardly a catastrophic number.

Nor is the United States alone in this trend toward net reproduction rates at one or below. As Figure 20 illustrates, intrinsic annual population growth rates have been dropping steadily among most of the industrialized nations, and it is highly probable that if 1973 data were available, practically all of the countries listed would be at net reproduction rates of about one.

Of course, it is true that the intrinsic annual population growth rates of the underdeveloped countries are still relatively high, but even there the population growth rates are trending steadily downward, and there is no reason to assume that they will not continue to do so. These nations, too, are improving their overall welfare.

In any event, it seems quite likely that net reproduction rates, which will generate zero population growth in the near future already have been reached in this country and in most of the industrialized countries on the basis of voluntary, private decisions. In my earlier discussion on the measurement of economic welfare, I emphasized sustainable per capita consumption, growing at the rate of technological progress, required

fifty-one

sufficient net investment to increase the capital stock at the natural rate of growth of the economy, taking into account both the increase in population and in productivity.

For a given capital output ratio, then, these lower rates of population increase mean that sustainable consumption per capita will be larger. Further, the capital widening requirement is decreased. In addition, if the rate of population growth is slowed, the equilibrium capital output ratio itself is changed. Average wealth of a population is a weighted average of the wealth positions of people of different ages. Over its life cycle the average family starts from a very low or even a negative net worth and then accumulates wealth to be spent in the years when children are going to school, houses are being built, and, of course, in old age. Consequently, a stationary or slow-growing population has quite a different wealth distribution from one that is growing rapidly. Exactly what this distribution will be is not immediately obvious, because, although there will be many fewer people in the early low-wealth years, there also will be more in the late low-wealth years.

Nordhaus and Tobin have estimated this shift with a series of studies and conclude that reduction in the rate of population growth increases the society's desired wealth-income ratio, which, in turn, means an increase in the capital output ratio. An increase in the capital output ratio will increase the society's sustainable consumption per capita. As the authors state:

> On both counts, therefore, a reduction in population increase should raise sustainable consumption. We have essayed an estimate of the magnitude of this gain. In a ZPG equilibrium sustainable consumption per capita would be 9-10 per cent higher than in a steady state of 2.1 per cent growth corresponding to 1960 fertility and mortality, and somewhat more than 3 per cent higher than in a steady state of 0.7 per cent growth corresponding to 1967 fertility and mortality.

> 'These neoclassical calculations do not take account of the lower pressure of population growth on natural resources. As between the 1960 equilibrium and ZPG, the diminished drag of resource limitations is worth about one-tenth of 1 per cent per annum in growth of per capita consumption. . . . .

* * *

Is growth obsolete? We think not. Although GNP and other national income aggregates are imperfect measures of welfare, the broad picture of secular progress which they convey remains after correction of their most obvious deficiencies. At present there is no reason to arrest general economic growth to conserve natural resources, although there is good reason to provide proper economic incentives to conserve resources which currently cost their users less than true social cost. Population growth cannot continue indefinitely, and evidently it is already slowing down in the United States. This slowdown will significantly increase sustainable per capita consumption. But even with ZPG there is no reason to shut off technological progress. The classical stationary state need not become our utopian norm.[49]

The Productive Society

II. Can We Have Growth and Freedom?

II. Can We Have Growth and Freedom?

We have satisfied ourselves reasonably well that there is a very strong correlation between a high-productivity society and real quality of life, and that in the United States neither natural resources nor overwhelming population growth is likely to limit our ability to improve our real economic welfare over the fore-seeable future. It now seems appropriate to examine the needs and pressures for growth both in the underdeveloped part of the world and here at home.

STRUCTURE OF THE WORLD ECONOMY, 1970

COUNTRY OR AREA	POPULATION (MILLIONS)	GNP (BILLIONS U.S. $ EQUIV.)	PERCENT OF WORLD POPULATION	PERCENT OF WORLD GNP	GNP PER CAPITA
INDUSTRIAL					
United States....	205	$ 977	5.6	27.5	$4,756
Canada........	21	87	0.6	2.4	4,047
Western Europe..	284	838	7.7	23.6	2,951
Japan.........	104	245	2.8	6.9	2,365
U.S.S.R.	243	486	6.6	13.6	2,000
Eastern Europe...	103	170	2.8	4.8	1,649
Others.........	18	54	0.5	1.5	2,956
Subtotal......	978	2,857	26.6	80.3	2,921
DEVELOPING					
India.........	580	53	15.8	1.5	91
China........	805	121	21.9	3.4	150
Others.........	1,311	528	35.7	14.9	403
Subtotal......	2,696	702	73.4	19.8	260
WORLD......	3,674	$3,560	100.0	100.0	$ 968

Figure 21[50]

Figure 21, with GNP values adjusted to give purchasing power equivalents (rather than simply using official exchange rates) illustrates the inevitable pressures strikingly. In 1970, the United States generated nearly 28% of the gross world product with 5.6% of the world's population. The total developed world generated more than 80% of the gross world product with less than 27% of the world's population. The 2.7 billion people in the underdeveloped world generated just under 20% of the gross world product with 73% of the world's total population.

The sheer scale of this need surely emphasizes the staggering task that faces the underdeveloped countries if they are to bring their material welfare up to any reasonable level. The combination of widespread education and almost instantaneous communications via radio and television has made the entire world conscious of the gap between the living standards in the industrialized countries and those in the developing countries. This, in turn, has been responsible for the concerns about limits to growth imposed by finite resources, by burgeoning populations, or both, and the expectations in the developing countries which are an ever-present political reality.

Figure 22, which I owe to Samuelson, shows that these facts are serious enough, but not catastrophic. In fact, the less-developed countries in recent years have had a considerably higher growth in real gross national product than have the developed countries, as the curves at left show. It is GNP per capita in the curves at right that really correlates with economic welfare, and the much higher rates of population growth in the less-developed countries have held down the GNP per capita there so that the gap between the less-developed and the developed countries in measurable economic welfare is increasing. Yet, the average GNP per capita growth of the LDC's has been very creditable, at approximately 3% for a half-dozen years, and not all that far behind that of the developed countries in 1972 and 1973.

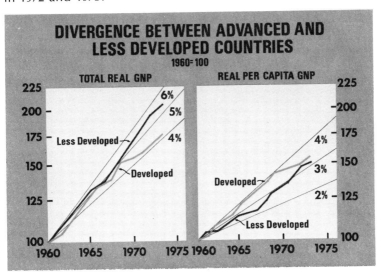

Figure 22[51]

Whatever the conclusions of the Forrester study, and in spite of the marked increase in anti-science and anti-technology statements in the United States, and in the rest of the developed world, the situation in these developing countries still is well-described in a statement by John Kenneth Galbraith, taken from a lecture he delivered in India in 1961 while he was U. S. Ambassador to that country:

... *The new countries of Asia and Africa are now concerned, as were those of Western Europe in the late eighteenth and early nineteenth centuries, to understand the processes on which progress depends. . .*

'Both in the new states and in the old states it has been recognized that economic development is an imperative. Indeed, this has been a distinguishing feature of the recent as compared with the earlier discussion. At least until the time of Marx, the problem of economic progress was explored with a measure of philosophical detachment. In the years since World War II it has been characterized by a note of high urgency. The nineteenth century discussion was in a world that was rather proud of what was happening. The twentieth century discussion is in a world which feels that a great deal more must happen and very soon.[52]

I would venture to say that, almost without exception, those who are most articulate in their denunciations of economic and material growth are enjoying family incomes in excess of $25,000 per year. Most would undoubtedly feel, and quite properly so, that the amenities they choose as representing quality of life for themselves in the way of education, standards of health, dress, residence, and general culture are not excessive, nor would they want to deny them to their fellow citizens. Yet, in 1971, only 6% of families in the United States had incomes in excess of $25,000 per year. Further, since just about half of the families in the country had incomes under $10,000 per year, and indeed nearly one-fifth under $5,000 per year,[53] it would appear clear that even in this most affluent of nations, we are far from providing sufficient growth to allow the majority of our citizens to choose these same amenities.

For most of the people of the world, the choice is clear. They will organize their societies to achieve what, in their eyes, we in the United States, Europe, and Japan already have. Further, here in the United States, the overwhelming majority of our citizens will not see themselves as having attained an affluence sufficient to accept, other than by force, mechanisms of political and social organization that would limit the future growth of our economy sufficiently to prevent their attaining an appreciably higher level of material welfare.

Thus, if the judgments I have formed from this collation of data and opinion are valid, the need and the pressure for growth are not only great but completely justified. Admitting the need for increased emphasis on quality of life, including protection of our environment, the key theoretical question and practical objective occupying leaders everywhere in the world, not the least here in the United States, remains: "How can we attain or sustain the productive society?"

For anyone of my generation (and my 60th birthday is not too many months away) the particular six decades through which we have lived assure that, for us, the analyses, judgments, and tenets of political economists are anything but abstract and academic.

From the shots fired in Sarajevo in June 1914 through the Great Depression of the 1930s, World War II, and the Cold War just now thawing, the headlines and the drama may have been seized by blood and the agony and the heroism of the battles on the land, on the sea, and in the air, or by the despair in the bread lines of the Thirties, or the horrors of Dachau, Buchenwald, or Katyn Forest. Yet, the real core of political struggle through the whole six decades has been about ways and means of achieving a productive society.

Karl Marx's prescriptions, as interpreted by Lenin and his successors, or by Mao, have led directly to those two opposite and opposing poles of communism, the Soviet Union and Red China.

We can't have forgotten in barely three decades that Hitler's Germany and Mussolini's Italy were both Machiavellian efforts at national socialism. And surely we remember it was Keynes' *General Theory* arising from his deliberations on the depression of the Thirties that is almost directly responsible for the economic practices of the mixed economies existing in a variety of forms throughout what we have grown to call the Free World and as especially exemplified by such countries as Western Germany, Sweden, France, Japan, and our own United States? Because all of these philosophical approaches to a productive society inherently find their expression in organizational forms, they lead to the generation of political systems. Where the emphasis has been overwhelmingly on the production and distribution of goods and services, the result has been the centralist and generally tyrannical, dominant state. Where the emphasis has been on the rights and responsibilities of the individual (And isn't it strange that this has occurred in the supposedly too materialistically oriented societies of the West and Japan?), the approach to centralism has been much more cautious. Yet, all of the statistics suggest that these have been the most productive societies of all.

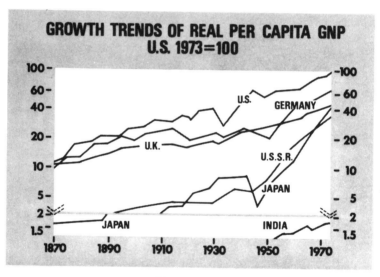

Figure 23[54]

Growth trends of real per capita GNP over the past century, as calculated by Samuelson, appear in Figure 23. Note especially the surging growth of Germany and Japan since 1950, but that the growth in GNP per capita in the Soviet Union also has been impressive. Undoubtedly, however, if one applied the Nordhaus and Tobin type of analysis to the Soviet Union to obtain a measure of economic welfare, it follows that their MEW would be still lower as compared to the U.S. because the USSR devotes a considerably smaller fraction of GNP to consumption. This is not too surprising when, with a GNP about half ours, the Soviet Union is spending just as much or even somewhat more on their military forces and space than we are, and it is certainly confirmed by the general observations of any visitor to Russia.

Samuelson suggests quite plausibly that the issue isn't so much the comparative strengths of the two economies now, but whether one is growing so much faster than the other as to be able to narrow or close the gap in the decades ahead.

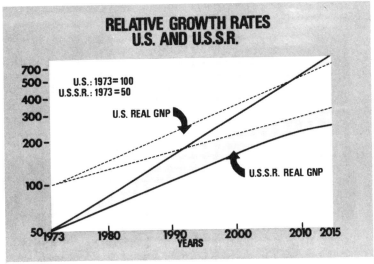

Figure 24[55]

In the data presented on Figure 24, Samuelson portrays the entire spread of growth rates estimated by experts for both the U.S. and the USSR for the next 40 years. The indices start with a U.S. base of real GNP in 1973 of 100 and a USSR base of 50. Using the highest rates estimated for both the USSR and the U.S., the USSR will cross the U.S. in real GNP sometime after the year 2000. On the other hand, if one takes the highest rate for the U.S. and the lowest rate estimated for the USSR, the United States would have a GNP more than three times as large as the Soviet Union 40 years from now instead of only twice as large, as in 1973. Obviously, then, it all depends on who is right.

Although estimates for the Soviet Union are hard to make, it does seem to be agreed that the USSR's post-World War II growth rates have been somewhat higher than ours, but they have been well below those of such mixed economies as West Germany and Japan. There is good reason to believe that it is easier for an economy that starts from a much lower level of productivity to show a higher initial rate of improvement. If so, as the Soviet Union reaches stages of development comparable to ours now and starts to make the shift to more emphasis on consumption and services, there is likely to be a retardation in her rate of growth.

Also, as the Soviet Union gets closer to the United States' GNP per capita, she also loses the advantage of simply copying more advanced technologies. These factors together are quite likely, I think, to militate in favor of the United States. Unless we flounder so in the decades ahead that we are unable to achieve the balance required for real growth in MEW, the gap a half century from now may well be sizable, and probably two to one or more.

As is well known, completely planned economies can have serious problems with productivity and growth. For example, Dr. Ota Sik, Deputy Premier of Czechoslovakia from April 1968 to August 1968 and head of that government's Commission for Economic Reform from 1963 to 1968, made a series of television talks to the citizens of Czechoslovakia in June and July of 1968, just before Soviet tanks came into Prague. He did so to acquaint the citizens of Czechoslovakia with their economy and to prepare them, not for a departure from Socialism, but for what he described as a more "flexible Socialist enterprise," one which very clearly would have given considerably more recognition to the market economy. In these discussions, he compared the Czechoslovakian economic performance controlled by Russian-style, highly centralized planning with that of a number of Western countries.

For example, Figure 25 compares the average per kilogram price of engineering exports for Czechoslovakia and West Germany over the 30-year span from 1936 to 1966.

Dr. Sik made these comments to accompany his data:

These figures show that Czechoslovakia engineering output is, in terms of its dollar-earning power, only half as effective as West Germany's and has even fallen below the level recorded for prewar Czechoslovakia . . . .

AVERAGE PER-KILOGRAM PRICE OF ENGINEERING EXPORTS ON CAPITALIST MARKETS EARNED BY CZECHOSLOVAKIA AND WEST GERMANY, 1936-1966 (IN U.S. DOLLARS)		
	WEST GERMANY	CZECHOSLOVAKIA
BEFORE WORLD WAR II	1.61	1.11
AFTER WORLD WAR II	1.44	1.87
1961	1.70	1.05
1963	1.79	1.11
1966	1.99	1.00

Figure 25[56]

'At bottom the key to our foreign trade lies in the abnormality of our economic relations. Whereas in every normally operating economy the full burden of business risk is borne by firms exposed to the pressure of a demanding world market, Czechoslovakia has allowed her socialist enterprises to exist apart from world competition.[57]

Dr. Sik used Figure 26 in his presentation showing Czechoslovakia's productivity about half that of Britain and West Germany and only one-quarter of that of the United States. On this issue of productivity, he commented as follows:

The . . . figures cited . . . demonstrate that the principles of management so stubbornly adhered to by the old guard among our politicians have failed to produce either an economic structure that meets the needs of the day or growth rate of labor productivity that can reduce the lead held by the Western world. . . .

'The only way out of this unhappy situation is to achieve a complete renaissance of the market, the instrument best fitted to gauge the success or failure of enterprises. Equally vital is scope for independent enterprise—after twenty years of taking orders, management teams will have to accustom themselves to the hitherto unknown feeling that, without orders, advice, or

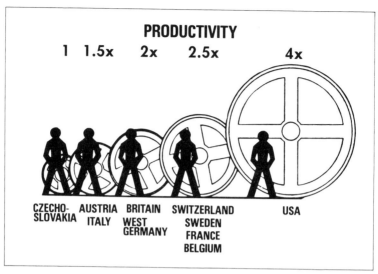

Figure 26[58]

approval from the top, they will be obliged to make full use of their own ideas. Indeed, with the market returning every failure, every lag behind world standards, with the precision of a boomerang, they will be forced to do so.[59]

Dr. Sik had a key observation as to why the economic enterprise of Czechoslovakia was laggard. He said:

Most important of all, the producer's interest in the market disappeared. It was bound to disappear when enterprise performance was completely divorced from the market. The enterprises had to go cap in hand to their superiors for everything they wanted, which in itself implied a total separation of production from market results. Consumer wants were really a matter of indifference—except to the consumers, of course. And we were all consumers.[60]

Remember that Dr. Sik, then Deputy Premier, made these presentations to the citizens of Czechoslovakia over their national TV network! Unfortunately, the Soviet Union did not feel it could afford to allow Czechoslovakia to deviate from the completely centrally controlled economy, Russian style, and so the tanks came in.

Since Dr. Sik was especially conscious of the superior performance of the German economy, as compared to that of Czechoslovakia, it seems appropriate now to review what Dr. Ludwig Erhard (vice chancellor and minister for economic affairs of the first West German Federal Republic, who led Germany to its economic miracle) stated in a book to the citizens of Germany to advise them with respect to German economic affairs:

What was the position when I was elected director of economic administration of the bi-zonal economic area on March 2, 1948? Much later—on May 31, 1954, in Antwerp—I described the period before currency reform in the following way:

'It was a time when most people did not want to believe that this experiment in currency and economic reform could succeed. It was a time when it was calculated that for every German there would be one plate every five years; a pair of shoes every twelve years; a suit every fifty years; that only every fifth infant would lie in its own napkins; and that only every third German would have a chance of being buried in his own coffin. That seemed to be the only life before us. This demonstrated the boundless delusion of planners that, on the basis of raw material stocks and other statistical data, the fate of a people could be determined for a long period in advance. These mechanists and dirigistes had absolutely no conception that, if a people were allowed once more to become aware of the value and worth of freedom, dynamic forces would be released.[61]

Actual production of the German bi-zone in 1947 amounted to only 39% of production in 1936. Contrast this with the West German situation a decade later, as described by Dr. Erhard.

The principle and application of what I have called 'die soziale Marktwirtschaft'—i. e., a social market economy—was given renewed support by a vast majority of people at the Bundestag elections of 15th September, 1957. Its success can be traced in every department of the country's economy and in every grade of society. I need only mention that our production today is nearly two and a half times that of the best pre-war year. Our industrial potential has multiplied six times since the currency reform. Our exports, estimated at DM 36 milliard (8.57 billion 1957 U. S. dollars) for the current year, put us in third place in world trade. Gold and foreign currency reserves have risen from nil in 1948 to today's total of DM 23 milliard (5.48 billion 1957 U.S.

dollars). It is especially gratifying to record that in the same period it was possible to treble the social services, as well as to build more than 4 million houses since 1949, the year when the first Federal Government came into operation.

'In outlining these events my sole concern is to demonstrate that we in West Germany have resorted to anything but a secret science. That, in fact, I have merely attempted to apply in practice the principles of modern national economics which all the countries of the West have evolved, with a view to overcoming the age-old antithesis of an unbridled liberalism and a soulless State control, to finding a sound middle way between out-and-out freedom and totalitarianism.[62]

With the success of mixed economies not just in generating overall growth in national product, but also in improving the economic welfare of the average citizen (and this is obvious even to the casual visitor to any of these countries), it would appear that the route to a higher and adequate economic welfare for people everywhere is clear. Just do what the mixed economies are doing. Use the market economy to the maximum extent possible and government intervention as necessary to assure full employment, control monopoly, and ensure justice, and all will be well. Surely, on the record, the completely centralized economies have not performed so well, and in the mixed economies the interventions by the state in the market system seem on the whole (but admitting some marked exceptions) to have improved the overall functioning of the system.

There are those who disagree. Friedrich Hayek warned that increasing intervention by government into the market economy was the inevitable "road to serfdom."

In August 1973 Paul Samuelson wrote in his *Newsweek* column:

. . . The old-time religion of a budget balanced at a low level, and of the government leaving key economic decisions to the marketplace, is no more the ideology of most Americans than is Billy Graham's version of religion the standard U. S. Code.

'Thirty years ago, Friedrich Hayek wrote 'The Road to Serfdom.' It was not a bad book, even though its principal thesis was wrong: Hitler's Fascism and Lenin's Communism were not, as Hayekians believe, the inevitable consequences of

Bismarckian-type social security, Lloyd George-type tax reforms or FDR New Deals.[63]

Unfortunately there are disquieting signs in the world, and I do not believe that I would want to dismiss Hayek's concerns as to the danger of increasing government control over the economy leading to intolerable encroachment on individual freedom in quite so cavalier a fashion as has Samuelson.

For example, the very ability of Allende to be elected in Chile with only 36% of the vote, stating as clearly as he did the avowed goal of turning Chile into a communist state, shows how readily it could happen.

Chile has run into real economic problems as a consequence of Allende's efforts. These problems were responsible for the recent military coup and Allende's alleged suicide. *The Economist*, in a very recent story on Chile after Allende, has the following to say:

> *The role of non-Chileans in training and organising the extreme left under Allende can hardly be exaggerated. The Tupamaros from Uruguay disposed of a rural base in the north of Chile, led by Sr Raul Bidegain Greissing, one of the few key leaders of the movement to escape the security forces in Montevideo. The Cubans were even more influential. They trained Dr. Allende's private guard at El Cañaveral and Sr Luis Fernandez de Oña, one of the men who organized Che Guevara's Bolivian expedition, regularly occupied a desk inside the Moneda palace after he married Allende's daughter Beatriz . . . .*
>
> *'The junta has still not made public the documents from the interior ministry which, it claims, contain evidence of plans for a left-wing coup which was supposed to begin with the assassination of senior officers and civilian leaders on September 17th. But over the past weeks other evidence has come to light that supports the idea that the extremists in the Allende government were preparing an insurrection of their own to 'complete the revolution.' The only doubt is whether Dr. Allende himself would have played the role of Lenin or Kerensky if that had finally come about.*[64]

Or consider the situation in Sweden, where the recent elections ended in a tie between the Socialist and non-Socialist bloc.

sixty-seven

Premier Olof Palme's Social Democratic Party lost six Parliamentary seats; his coalition partners, the Communists, were the big winners, and their gains produced the tie. This is in a country in which the mixed economy, many believe, has been the most effective of all and which has generated a GNP per capita in 1973—higher than that of the United States.

Yet, there have been some strange currents in Sweden in recent years. There have been many observations about the increasing anti-Americanism there. Visitors and foreigners living in Sweden have commented that the national network television news programs have emphasized rather persistently apparent American, British, or French imperialism and have de-emphasized such items as the increasing struggles for personal freedom inside the Soviet Union or the use of troops in Czechoslovakia to overturn the economic reform taking place there in 1968. The tax burden has become enormous. About 42% of Sweden's GNP is funneled into taxes, as compared with 27% in the U.S. Of every dollar a Swede earns above $12,000 per year, 70% goes into taxes.[65] Because growth has been somewhat more limited the last few years in Sweden, unemployment has increased; and although at 3.6%, low in U. S. terms, it is one of the highest in Europe. Inflation, too, has been a problem in Sweden, as it has been in all of the mixed economies, and the Social Democrats have suggested programs to combat inflation which display clearly how readily much more government encroachment could take place, particularly now that the Communists are playing a deciding role.

Here is what *The Economist* had to say last August on this subject before the Swedish elections:

The Social Democrats now say they will reduce inflation by more interventionism. Their most controversial move is that the huge state pension fund is to be given power to buy equity shares.

The Swedish state pension fund now has assets of £5.5 billion (13.4 billion dollars) compared with £2.3 billion (5.6 billion dollars) in 1968 and a projected £15.4 billion (37.5 billion dollars) in 1980. The inflow accounts for half of all the new insurance-type savings in the country, and by 1978 the assets will exceed those of all the banks, insurance and credit institutions. From next January the fund will be allowed to invest £48m (116.9 million dollars) in equities with more later as the government decides. Not too much for a start, but with no limit on the amount that may be put into any one stock. Business and the opposition attack this as the thin end of a very thick wedge that could lead to unlimited back-door nationalisation.[66]

Or consider what has been happening in Canada as reported by John Chamberlain in a recent newspaper story:

. . . You would not normally expect to see the idea of just compensation flouted in Canada, a country with the same legal traditions as the U.S. . . . .

'A US-owned company, Churchill Forest Industries, made a deal some years ago with what was then the Conservative government of the Canadian province of Manitoba to develop some 40,000 square miles of country in the northern part of the province . . . Manitoba's share of the financing came to 86% of the total, with Churchill Forest Industries putting up the remainder . . . Churchill was to build the necessary mills, lay out new towns for mill workers and their families, and market the forest products for a period of 60 years . . . .

'This was before the New Democratic Party, elected on a socialistic nationalization of resources platform, took control of Manitoba in 1969. What followed has been Allendeism pure and simple. Edward Schreyer, the New Democratic Party premier of Manitoba, has simply ousted Churchill Forest Industries without paying a nickel for Churchill's rights in the original deal. The money that Churchill spent in completing two sawmills, one paper mill and one pulp mill is apparently going for naught. The big international copper companies, Kennecott and Anaconda, got no worse treatment in Chile.

'Churchill, unable to press its suit in Manitoba, is trying to sue for retribution in an American state court, citing damage and compensation claims of more than $500 million. The New Jersey Superior Court, though recognizing the "raging Canadian political and legal turmoil" that produced the takeover, has denied Churchill any use of New Jersey courts on the ground that local US courts don't provide the right locale for international compensation appeals . . . Manitoba did not deny the expropriation: it simply said, in effect, that Churchill isn't entitled to any compensation arranged for by an ousted Conservative government that made deals with "mongrel" capitalists.[67]

sixty-nine

Even these few examples are sufficient to demonstrate that one cannot simply dismiss Hayek's concerns as groundless. The very success of the mixed economies, making use as they do of government intervention, can lull or confuse their citizenry to the point where they not only allow but may even demand the kind of government intervention that will eventually destroy their personal freedoms as well as the ability of the society to perform well economically.

Of course, there are those who simply question the meaning of such values as individual responsibility, personal freedom, or human dignity, and from that conclude that concern about these values should not play a significant role in determining the manner in which society is organized.

Psychologist B. F. Skinner, for example, in his immensely successful book, *Beyond Freedom and Dignity*, published in 1971, puts it as follows:

Freedom and dignity illustrate the difficulty. They are the possessions of the autonomous man of traditional theory, and they are essential to practices in which a person is held responsible for his conduct and given credit for his achievements. A scientific analysis shifts both the responsibility and the achievement to the environment. It also raises questions concerning 'values.' Who will use a technology and to what ends? Until these issues are resolved, a technology of behavior will continue to be rejected, and with it possibly the only way to solve our problems.

'Man's struggle for freedom is not due to a will to be free, but to certain behavioral processes characteristic of the human organism, the chief effect of which is the avoidance of or escape from so-called 'aversive' features of the environment. Physical and biological technologies have been mainly concerned with natural aversive stimuli; the struggle for freedom is concerned with stimuli intentionally arranged by other people. The literature of freedom has identified the other people and has proposed ways of escaping from them or weakening or destroying their power. It has been successful in reducing the aversive stimuli used in intentional control, but it has made the mistake of defining freedom in terms of states of mind or feelings, and it has therefore not been able to deal effectively with techniques of control which do not breed escape or revolt but nevertheless

have aversive consequences. It has been forced to brand all control as wrong and to misrepresent many of the advantages to be gained from a social environment. It is unprepared for the next step, which is not to free men from control but to analyze and change the kinds of control to which they are exposed.[68]

Or consider this dialogue from Skinner's *Walden Two*. He has Frazier, who is the creator and developer of Walden Two, say:

Democracy is not a guarantee against despotism, Burris. Its virtues are of another sort. It has proved itself clearly superior to the despotic rule of a small elite. We have seen it survive in conflict with the despotic pattern in World War II. The democratic peoples proved themselves superior just because of their democracy. . . .

'But the triumph of democracy doesn't mean it's the best government. It was merely the better in a contest with a conspicuously bad one. Let's not stop with democracy. It isn't, and can't be, the best form of government, because it's based on a scientifically invalid conception of man. It fails to take account of the fact that in the long run man is determined by the state. A laissez-faire *philosophy which trusts to the inherent goodness and wisdom of the common man is incompatible with the observed fact that men are made good or bad and wise or foolish by the environment in which they grow.*

* * *

Dictatorship and freedom—predestination and free will,' Frazier continued, 'What are these but pseudo-questions of linguistic origin? When we ask what Man can make of Man, we don't mean the same thing by "Man" in both instances. We mean to ask what a few men can make of mankind. And that's the all-absorbing question of the twentieth century. What kind of world can we build—those of us who understand the science of behavior?'

'Then Castle was right. You're a dictator, after all.'

'No more than God. Or rather less so. Generally, I've let things alone. I've never stepped in to wipe out the evil works of men with a great flood. Nor have I sent a personal emissary to reveal my plan and to put my people back on the track. The original design took deviations into account and provided automatic corrections. It's rather an improvement upon Genesis.[69]

seventy-one

Skinner thus denigrates the conception of freedom into simply a consequence of environment.

It is true that Walden Two appears to be a benevolent society, but I am struck by the very different consequences envisioned by Orwell in his *1984* following from not dissimilar attitudes. Here are Orwell's words, as expressed by O'Brien in the brainwashing of Orwell's central character, Winston:

We are the priests of power,' he said. 'God is power. But at present power is only a word so far as you are concerned. It is time for you to gather some idea of what power means. The first thing you must realize is that power is collective. The individual only has power in so far as he ceases to be an individual. You know the Party slogan "Freedom is Slavery." Has it ever occurred to you that it is reversible? Slavery is freedom. Alone—free—the human being is always defeated. It must be so, because every human being is doomed to die, which is the greatest of all failures. But if he can make complete, utter submission, if he can escape from his identity, if he can merge himself in the Party so that he is the Party, then he is all-powerful and immortal. The second thing for you to realize is that power is power over human beings. Over the body—but, above all, over the mind. Power over matter—external reality, as you would call it—is not important. Already our control over matter is absolute.

* * *

Winston shrank back upon the bed. Whatever he said, the swift answer crushed him like a bludgeon. And yet he knew, he knew, that he was in the right. The belief that nothing exists outside your own mind—surely there must be some way of demonstrating that it was false. Had it not been exposed long ago as a fallacy? There was even a name for it, which he had forgotten. A faint smile twitched the corners of O'Brien's mouth as he looked down at him.

'I told you, Winston,' he said, 'that metaphysics is not your strong point. The word you are trying to think of is solipsism. But you are mistaken. This is not solipsism. Collective solipsism, if you like. But that is a different thing; in fact, the opposite thing. All this is a disgression,' he added in a different tone. 'The real power, the power we have to fight for night and day, is not power over things, but over men.' He paused, and for a moment *assumed again his air of a schoolmaster questioning a promising pupil: 'How does one man assert his power over another, Winston?'*

Winston thought. 'By making him suffer', he said.[70]

My own bias, I suppose, is clear just from the manner in which I have organized this presentation. Personal responsibility, liberty, human dignity, are values I consider of grave importance in any society that hopes to provide a life of quality, not just quantity.

My convictions can't help but be strengthened by the impassioned plea of the Soviet novelist, Aleksandr Solzhenitsyn, in which he nominated Andrei Sakharov for the 1973 Nobel Peace Prize for "Sakharov's untiring and sacrificing (and personally dangerous) resistance to unceasing state violence against individual and national population groups." Here are some excerpts from that letter:

One example—the terrorism of the past few years. While man is tense and on his guard against wars, he has a tendency to fail to detect other forms of violence. The confusion is complete, and people are not prepared to reject terrorism committed by a single, little, individual. And most astonishing, a world-wide humanitarian organization is incapable of securing a moral condemnation of terrorism!

'One could jokingly suggest the following: 'When we are attacked, that is terrorism. But when we are the attacker, then it's a partisan freedom movement.

'Permanent state violence—which throughout the decades it has reigned has succeeded in taking over all 'judicial' forms, codifying thick collections of its violent 'laws,' draping capes across the shoulders of its 'judges'—is the most threatening danger in our world of today, even if it is only barely recognized or understood.

'This violence no longer needs to place explosives under something or to toss bombs. Its procedures are carried out in strict silence, seldom disturbed by the final shrieks of he who is being strangled. This type of violence permits itself to take on a respectable appearance.

'There is an emotional error involved in the comprehension of what is included in the conception of "peace."

seventy-three

'We do not err because it may be difficult to see the truth floating on the surface. But we err because it is pleasant and easy to seek an understanding in exact conformity with our feelings—especially our egotistical feelings. The truth has been around a long time. It has been exhibited, proven and explained. But it draws no attention or sympathy, just as in the case of Orwell's '1984,' with its 'conspiracy of flattery.'

* * *

I have throughout my years devoted myself to examining Russian life before it was ruined, and I have been personally impressed by the apparently impossible resemblance between the Russian czarist regime in its final years and, for example, the Republic of the United States of recent years—years, I venture to predict, that also are the final ones before the great chaos. This is not a comparison of material and economic qualities and not a comparison of social structures, but of something more important: a comparison of the politicians' lack of emotional ability to reflect. The entire storm of wrath of the Democrats on the subject of Watergate resembles a parody of the angry, thoughtless storm of the Cadets against Goremykin-Sturmer in 1915-1916.

* * *

On the other hand, one must agree—as so many, many maintain—that what is happening in the Soviet Union is not just something in 'just about every country' but is the tomorrow of mankind and is thus, in the matter of its inner processes, worth full attention by Western observers.

'No, it's not the troubles involved with gaining an insight that pose difficulties for the West. Rather, it's the lack of a desire to know, the emotional preference of the comfortable solution instead of the difficult one. Such a searching for insight is fed by the Munich Spirit, by concession and the spirit of compromise, led by an anxious self-deception on the part of societies of good intentions and persons who have lost their determination to make sacrifices and stand firm.[71]

So today's Solzhenitsyn joins the warnings of Hayek thirty years ago. How else could what is happening in the Soviet Union be the "tomorrow of Mankind?"

Earlier, in connection with the report on Sweden from *The Economist*, I commented that inflation was indeed a serious problem in the mixed economies. Wilhelm Röpke, an economist who helped guide the post-war economic recovery of West Germany, discussing the influence of John Maynard Keynes, observed as follows:

He remained a liberal, professing devotion to democratic freedoms and convinced that in his singular way, he was promoting them. Another constant in his career was his belief, derived from his expanding researches in monetary theory, that the real defect of our economic system must be sought in the organization of its finances and its monetary institutions. To improve this organization, he made proposals ranging from the moderate ones of his Tract on Monetary Reform *(1923) to the radical ones of his last great work,* The General Theory of Employment, Interest and Money *(1936).*

'This is not the place to evaluate in detail the services which Keynes, in these works, rendered to the advancement of theory. Unquestionably, they are considerable. At the same time, it is precisely because he so deeply influenced his time that it is necessary to ask whether the practical results of his theories and proposals, which were intended to improve the working of the existing economic system, did not ultimately have the effect of weakening its foundations—so that Keynes, in tragic opposition to his own intention, must be numbered among the gravediggers of that very order of liberal democracy to which his innermost allegiance belonged.

'One may believe that there are times in which vigorous measures to increase the money supply will prevent disaster; but not with impunity can a leading scientific figure like Keynes bestow the mantle of his authority on the chronic propensity of all governments to inflate. [72]

That inflation is a key, perhaps *the* key, problem faced by the mixed economies hardly can be questioned when the increase in consumer prices over the past year in these countries is examined.

I presume one might debate the presence of Yugoslavia in Figure 27's list of mixed economies, but since there is some attempt in that country to depend upon the market and the price system, I have included it. You will note that over the past year consumer prices have increased from the United States' 5.7%, through Denmark's 8.8%, up to Yugoslavia's nearly 20%.

seventy-five

PERCENT INCREASE IN CONSUMER PRICES
JULY 1972 – JULY 1973

Yugoslavia	19.7	Netherlands	8.4
Greece	13.1	Switzerland	8.3
Finland	12.1	Australia	8.2
Spain	12.1	France	7.4
Portugal	12.0	Norway	7.4
Japan	11.9	Germany	7.2
Italy	11.8	Austria	7.0
Ireland	11.7	Belgium	6.6
Britain	9.4	Sweden	6.6
Denmark	8.8	UNITED STATES	5.7

Figure 27[73]

I have included this list not merely to emphasize the severity of inflation in the mixed economies but to remind us that experience has demonstrated the extraordinary difficulty of halting it through government intervention.

Illustrative of the danger are a few paragraphs reconstructed from an address by Economist Joseph Schumpeter in 1949:

I do not pretend to prophesy; I merely recognize the facts and point out the tendencies which those facts indicate.

'Perennial inflationary pressure can play an important part in the eventual conquest of the private-enterprise system by the bureaucracy—the resultant frictions and deadlocks being attributed to private enterprise and used as arguments for further restrictions and regulations. I do not say that any group follows this line with conscious purpose, but purposes are never wholly conscious. A situation may well emerge in which most people will consider complete planning as the smallest of possible evils. They will certainly not call it socialism or communism, and presumably they will make some exceptions for the farmer, the retailer and the small producer; under these circumstances, capitalism (the free-enterprise system) as a scheme of values, a way of life, and a civilization may not be worth bothering about.

'Whether the American genius for mass production, on whose past performance all optimism for this way of life rests, is up to this test, I dare not affirm; nor do I dare to affirm that the policies responsible for this situation might be reversed.

'Marx was wrong in his diagnosis of the manner in which capitalist society would break down; he was not wrong in the prediction that it would break down eventually. The Stagnationists are wrong in their diagnosis of the reasons why the capitalist process should stagnate; they may still turn out to be right in their prognosis that it will stagnate—with sufficient help from the public sector.[74]

Certainly we have had the spectacle of a Republican administration, inherently opposed to and uncomfortable with wage and price controls, applying them in this country with, at best, limited success, yet undoubtedly doing so because it was felt to be politically necessary. In addition, in the spring of 1971, prior to the initial application of the controls in this country, the majority of businessmen with whom I talked expressed themselves as feeling that wage controls were a necessity. Yet, wage controls without price controls were obviously a political absurdity. Now that controls have operated in this country for something more than two years, the overwhelming majority of the businessmen with whom I talk are equally convinced that the controls were a mistake and should be lifted just as soon as possible.

This is what Dr. C. Jackson Grayson, who as chairman of the Price Commission in Phase 2 had very recent experience on which to base his judgment, said about price controls:

Almost every time we tried to adjust our economic system to correct one problem, two or three more were created, and the more we felt the temptation to 'control.'

'In the end, I believe that any prolonged control system would disrupt the free market system.

* * *

Grayson pointed to three trends he believes have contributed to the movement toward public control.

Business and labor too often seek to reduce rather than to encourage competition in their markets.

'Continuing price and wage controls are leading the public to believe that central planning and control are superior, mandatory and desirable.

'Americans, in distrusting the market system, are demanding more economic benefits from the Federal government and are seeking ways to insulate themselves from the impact of economic change.

'The best solution to the problems of the economy will come in better functioning of the private competitive system and a better quality, not quantity, of public control.[75]

In spite of the dangers (and they are real as these examples illustrate) that government intervention in the mixed economy will eventually lead to overcentralism and finally become the road to serfdom, I see no alternative to a strong role for government in today's world. The real dilemma faced by men everywhere in world society is that their needs are so great that they really have no choice except to opt for modes of organization of society that assure high productivity. Further, while an organization of society that encroaches on personal freedom will be tolerated, one that cannot provide an adequate and increasingly higher level of material existence, adequate education, good health, and long life will not be tolerated if it is perceived as such.

To the extent that the warnings with which I began these discussions alert us to the ways in which a blind and unthinking pursuit of growth can be counter-productive and depress our quality of life, they are useful. But when such critics see the only route to the alleviation of these pressures as the abandonment of growth, their recommendations become counter-productive and simply will not have a significant influence on the choices of the great masses of population of the world.

Additionally, there is the very real danger that the emotional appeal of these criticisms will result in irrational extensions of centralized government, introduce unnecessary restrictions on individual and local freedom, and finally so enlarge the regime of regulation with one regulation leading to another and another as to thrust us all unwittingly into Solzhenitsyn's "tomorrow of mankind."

I don't believe it has to happen if we choose wisely enough from among the variety of routes that a strong role for government can follow. Here once again I would like to turn to Wilhelm Röpke. Here is how he saw it:

> *Order and incentive in the economy—these, then, are the two cardinal problems around which everything resolves and which from minute to minute must be freely and noiselessly solved. We find, however, if we extend our inquiry back to first principles, that there are only two possible solutions to these problems (if we exclude the special case of the self-sufficient peasant economy). The two possible solutions, as we know already, are those of freedom and command. That of freedom means the strict adherence to an order functioning with astonishing regularity through the medium of the free market with its freely fluctuating prices. That of command, however, means an economic order in which order and incentive are placed in the hands of the consciously ordering, planning, inciting, commanding, and command-enforcing state. The one we call market economy, the other command economy, planned economy, centrally administered economy, collectivist (socialist) economy. It cannot be too strongly emphasized that as far as the task of ordering economic life is concerned, we have only this exclusive choice between market economy and command economy. We cannot take refuge in some third alternative, in cooperatives, trade unions, in undertakings patterned after the much-cited but much misunderstood Tennessee Valley Authority, corporatism, industry council plans, vocational orders, or any other form of 'ersatz' socialism. We must choose between price or state command, between the market or the authorities, between economic freedom or bureaucracy. Having tried out both systems, however, we know only too well that, in fact, we have no further choice in the matter. It has been shown that Western man is not free to opt for a collectivist system, since the latter is unable to guarantee an effective system of order and incentives which would be compatible with freedom and with the existence of an international community. He who chooses the market economy must, however, also choose: free formation of prices, competition, risk of loss and chance for gain, individual responsibility, free enterprise, private property.*
>
> *'This choice—and herewith we return to the heart of our argument—has nothing whatsoever to do with what was formerly understood under the terms 'free economy' or 'capital-*

seventy-nine

ism.' The new orientation of economic policy—along a path which the author has designated as the 'third road'—consists precisely in this: that we recognize the impassibility of the socialist road without our feeling it necessary, on that account, to return to the old worn-out road of 'capitalism.' Two important planks must be included in this program whose aim it is to ensure the existence of a natural order. The first calls for a stable framework which, as already observed, is indispensable to a well-ordered market economy. This in itself means that the state has a number of important tasks to fulfill: the establishment of a healthy money system and a prudent credit policy which together will serve to eliminate an important source of economic disturbances. Such a framework will also necessitate a legal system carefully constructed to prevent, as far as possible, abuse of the freedom of the market and to ensure that the road to success will be entered only through the small door of reciprocal service. In a word, this framework should be designed so as to reduce to a minimum the numerous imperfections of the market economy.[76]

I believe that Röpke has stated well the principles that must guide the inevitable government interventions. Unfortunately, determining exactly how to apply these principles is not easy, because even well-qualified economists who would accept these overall principles disagree sharply on mechanisms of government intervention, especially as to what is a framework designed to reduce to a minimum the numerous imperfections of the market; and politically, it is often difficult to stick to them firmly.

It will help us to understand why it will be difficult to assure only appropriate government interventions if we review a few of the key changes in our economy over the past century and remind ourselves that economies based upon a division of labor and organization are anything but recent in origin.

Plato's straightforward logic and irresistible conclusions in his "Discourses on The Republic" make obvious not only that man has understood for more than 2400 years the economic gains a division of labor would bring but also that the thrust from agriculture to the city and the desire to cultivate the kind of life the city makes possible is anything but a contemporary development.

Now the first and greatest of necessities is food, which is the condition of life and existence.

Certainly.

The second is a dwelling, and the third clothing and the like.

True.

And now let us see how our city will be able to supply this great demand: We may suppose that one man is a husbandman, another a builder, some one else a weaver—shall we add to them a shoemaker, or perhaps some other purveyor to our bodily wants?

Quite right.

The barest notion of a State must include four or five men.

Clearly.

And how will they proceed? Will each bring the result of his labours into a common stock?—the individual husbandman, for example, producing for four, and labouring four times as long and as much as he need in the provision of food with which he supplies others as well as himself; or will he have nothing to do with others and not be at the trouble of producing for them, but provide for himself alone a fourth of the food in a fourth of the time, and in the remaining three-fourths of his time be employed in making a house or a coat or a pair of shoes, having no partnership with others, but supplying himself all his own wants?

Adeimantus thought that he should aim at producing food only and not at producing everything.

Probably, I replied, that would be the better way; and when I hear you say this, I am myself reminded that we are not all alike; there are diversities of natures among us which are adapted to different occupations.

Very true.

And will you have a work better done when the workman has many occupations, or when he has only one?

When he has only one.

* * *

And if so, we must infer that all things are produced more plentifully and easily and of a better quality when one man does one thing which is natural to him and does it at the right time, and leaves other things.

Undoubtedly.

eighty-one

Then more than four citizens will be required; for the husbandman will not make his own plough or mattock, or other implements of agriculture, if they are to be good for anything. Neither will the builder make his tools—and he too needs many; and in like manner the weaver and shoemaker.

True.

* * *

Then there must be another class of citizens who will bring the required supply from another city?

There must.

But if the trader goes empty-handed, having nothing which they require who would supply his need, he will come back empty-handed.

That is certain.

And therefore what they produce at home must be not only enough for themselves, but such both in quantity and quality as to accommodate those from whom their wants are supplied.

Very true.

Then more husbandmen and more artisans will be required?

They will.

Not to mention the importers and exporters, who are called merchants?

Yes.

* * *

And if merchandise is to be carried over the sea, skilful sailors will also be needed, and in considerable numbers?

Yes, in considerable numbers.

Then, again, within the city, how will they exchange their productions? To secure such an exchange was, as you will remember, one of our principal objects when we formed them into a society and constituted a State.

Clearly they will buy and sell.

Then they will need a market-place, and a money-token for purposes of exchange.

* * *

True.

* * *

Yes, Socrates, he said, and if you were providing for a city of pigs, how else would you feed the beasts?

But what would you have, Glaucon? I replied.

Why, he said, you should give them the ordinary conveniences of life. People who are to be comfortable are accustomed to lie on sofas, and dine off tables, and they should have sauces and sweets in the modern style.

Yes, I said, now I understand: the question which you would have me consider is, not only how a State, but how a luxurious State is created; and possibly there is no harm in this, for in such a State we shall be more likely to see how justice and injustice originate. In my opinion the true and healthy constitution of the State is the one which I have described. But if you wish also to see a State at fever-heat, I have no objection. For I suspect that many will not be satisfied with the simpler way of life. They will be for adding sofas, and tables, and other furniture; also dainties, and perfumes, and incense, and courtesans, and cakes, all these not of one sort only, but in every variety; we must go beyond the necessaries of which I was at first speaking, such as houses, and clothes, and shoes: the arts of the painter and the embroiderer will have to be set in motion, and gold and ivory and all sorts of materials must be procured.

True he said. [77]

Plato has described not just the needs of men, but the desires of men with respect to how they want to organize their lives. Yet the facts are that until a bare 200 years ago, it took all but a handful of the working populations to provide his "first and greatest of necessities—food."

It is almost impossible for me to appreciate fully that in spite of this clear comprehension developed and understood as early as Plato 2400 years ago, almost everywhere the division of labor proceeded so slowly that even 200 years ago only a handful of men were governors, craftsmen, retailers, artists, scholars, merchants, traders, and priests—the overwhelming majority were still engaged in producing food. [78]

Probably up to the late 1700s, there was no settled community in which 90% or more of the community were not engaged in tilling the soil. [79] It took that long before there was sufficient technology to provide the means that would allow men even that much division of labor and enough improvement in productivity to begin to generate significant material wealth for more than a handful of the State's citizens.

The new heavy plow, the horseshoe and the padded collar-

type harness are all important inventions, but it wasn't until sometime after the year 1000 that the heavy plow and horsepower were finally combined and thus increased the productivity in agriculture. It also accelerated the movement of which we are so conscious today from the farm to the city. The ox moved very slowly, and the peasant had to live close to his fields. The horse, moving much more rapidly, allowed the peasant to travel a much greater distance and extensive regions "once scattered with tiny hamlets, came to be cultivated wildernesses dominated by huge villages which remained economically agrarian, for the most part, but which in architecture and even in mode of life became astonishingly urban."[80]

The point is that man, from nearly the beginning of recorded history, has sought out technology to improve his productivity and his way of life, and he has moved from the village to the town to the city because he wanted to and as his increasing productivity allowed him to do so.

Of course, all over the world, man is continuing that transition.

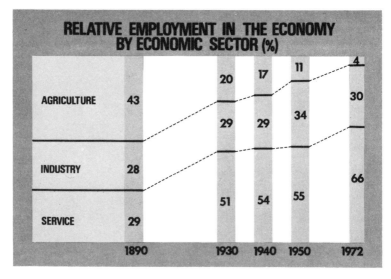

Figure 28[81]

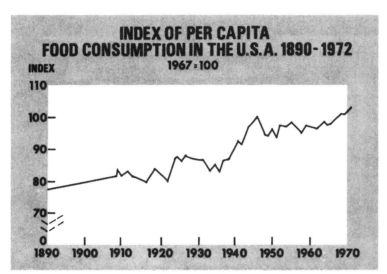

Figure 29[82]

Here in the United States, as recently as 1890, 43% of our working population was still engaged in agriculture (Figure 28*), but this was still probably about half of what it had taken not much more than a century before. The improvement in productivity is what allowed the development of industry, the increase in services, and the move to the cities. Within 40 years, by 1930, only 20% of our working population was required on the farms. That transition has continued, and in 1972, a bare 4% of our working population produced a wider variety and an increasing quantity of food for all of us and for export.

The degree to which this is true is portrayed in Figure 29. It shows that while this enormous increase in productivity was going on, per capita food consumption in the United States increased nearly 35%.

We had an estimated 9.4 million people in agriculture in 1890 to produce the food and other raw agricultural products necessary for a population of 63.1 million. In 1972, only 3.3 million

---

*Included in agricultural employment are those persons engaged in farms, agricultural services, forestry and fisheries. The industry sector is composed of persons engaged in mining, contract construction and manufacturing. Persons engaged in the service sector include those in transportation, communications, utilities, wholesale and retail trade, finance, insurance and real estate, services and government.

(35% of the 1890 figure) were required to produce the agricultural products for a population of 208.8 million. Indeed, had we been producing in 1972 as we were in 1890, it would have taken 41.6 million workers to produce the quantity of agricultural products consumed and exported in 1972. Since only 3.3 million actually were so engaged, a total of 38.3 million workers were released to man our factories and produce the industrial goods; but especially to move into transportation, communication, the public utilities, into trade and finance and insurance and real estate, into household employment, into health and education, and into government in general.

This very striking increase in productivity per person came about because of improved technology showing up in such specific ways as improved seed grains, improved livestock, increased yields per acre through proper fertilizing, better land cultivation and preparation, including irrigation, and markedly improved tools, all the way from seeding and preparation to cultivation and harvesting.

To take advantage of these gains, a more knowledgeable work force was required. Men and women engaged in agriculture today are better educated than they were in 1890. Their farms are larger, and they have invested much more heavily in the capital goods and technology to farm at this more productive level.

Of course, it took workers to generate the knowledge that led to the gains to improve the seed and livestock, to design and build the tools, to educate the workers in agriculture and all of the others who contributed to the agricultural gains, or to make butter, cream, and cheese in factories instead of on the farm as it once was done; but this is typical of the substitution occurring through education and technology together with the consequent product and productivity gains.

But the shift from agriculture to industry is not the only shift. The United States has been described as an industrial society. Yet, the proportion of our working force in industry, in spite of the enormous increase in variety and quality of product, has remained remarkably constant—28% back in 1890, reaching a peak of about 34% in 1950, and decreasing to just 30% in 1972. The facts are then that we are able in the United States to produce all of our food and goods with barely one-third of our working population. The remaining 66% are engaged in the service sector. Part of that shift to services is through necessity. We need the transportation and the communication, trade and finance to allow the division of labor in agriculture and industry. Moving into the cities means that we need police forces and more complex governments, water, sanitation, and the multitude of other special services it takes just to make a modern city possible.

But the shift has occurred also because we wanted it to. We want the amenities provided by urban life. We want to be able to congregate. We want the educational institutions it makes possible. We prefer to work in ways that decrease physical labor. We want the increased variety of foods and goods that congregating in the large communities makes possible. Furthermore, this shift is going to go on.

Figure 30 shows the relative gross national product originating in agriculture, industry and services. In 1950, 11% of the working population was engaged in agriculture. Note that 11% produced only 4.9% of the gross national product in 1950. In 1972 only 4% of the working population was producing 2.8% of the gross national product. Although a decreased percentage in 1971 dollars, the value of the gross national product originating

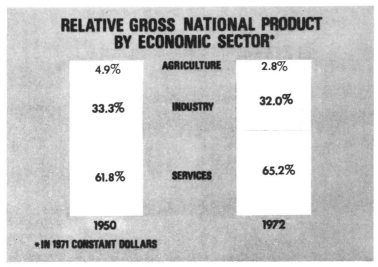

Figure 30[83]

in the agricultural section increased from $25 billion in 1950 to $32 billion in 1972, yet required a much smaller percentage of the work force.

Note that there was also a significant gain in industry. In 1950, 34% of the working population produced 33.3% of the gross national product. In 1972, only 30% of the working population was producing an increased 32% of the gross national product. The absolute values in 1971 dollars had more than doubled, from $168 billion to $355 billion.

In the service sector, on the other hand, although between 1950 and 1972 the percentage of the work force increased 11%, from 55% to 66%, the percentage of GNP originating increased only from 62% to about 65%. This is at least partly because in the increasing government sector the GNP originating is expressed in terms of the input dollars (i.e., the dollars expended on costs) because of the difficulty of arriving at product originating. But it is also at least partially because of inability to apply technology and capital to improve productivity in the services area to anything like the same extent we have been able to do in agriculture and industry.

Figure 31 displays in somewhat more detail the proportion of the service sector made up by workers employed directly by government:

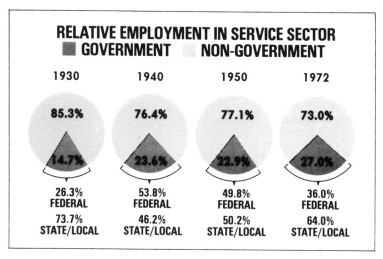

RELATIVE EMPLOYMENT IN SERVICE SECTOR
■ GOVERNMENT   NON-GOVERNMENT

1930	1940	1950	1972
85.3%	76.4%	77.1%	73.0%
14.7%	23.6%	22.9%	27.0%
26.3% FEDERAL	53.8% FEDERAL	49.8% FEDERAL	36.0% FEDERAL
73.7% STATE/LOCAL	46.2% STATE/LOCAL	50.2% STATE/LOCAL	64.0% STATE/LOCAL

Figure 31[84]

The percentage of total services in government increased from just about 15% to 27% in the 42 years. Because the service sector itself had grown from 29% to 66% of the working population over the same years, the absolute change is startling. We had 3.3 million people in government in 1930. In 1972 we had almost 15 million.

There is a second phenomenon, of which perhaps we are not all as aware as we should be. We are inclined to be especially querulous about the increase in employment in our Federal Government (and we need to be because the concerns about the loss of freedoms are probably quite properly focused on increases there rather than in the local areas). By 1972, although the absolute numbers in Federal Government had increased from 871,000 in 1930 to 3.3 million in 1940 and 5.2 million in 1972, the facts are that Federal employment was down to 36% of total government employment. In 1944, during World War II, because of the large number in the Armed Services, the Federal portion of total government reached as high as 84%. State and local government, which in 1930 was nearly 74%, decreased by 1940 to 46.2%, and by 1972 grew to 64% of total government with a total of 9,250,000 employed. Nearly half of these workers in state and local government, or 4.5 million, were in public education.

I would like to re-emphasize that this increase in the service sector, including that in government, has come into being because, as citizens, we wanted the services provided. The increasing productivity of the agricultural and industrial sector allowed us to make the shift and also improved our standard of living.

It also brought along some negatives. The shift from the farms to the cities produced the congestion in our cities and the ghettos. The technology that increased the productivity of the acreage contributed to the pollution of our streams with the residue of chemical fertilizers. Further, while the overwhelming majority of the workers who came from the farms to man jobs in industry and in services were so competent that they played a principal part not just in manning, but in managing, the institutions generated, there was a minority of workers, untrained and maladjusted, who moved into the crowded city ghettos and were unable to obtain work they were capable of performing.

There are two other key effects. First of all, we do not know how to improve productivity in much of the services sector to anything like the same extent we do in agriculture and industry. One merely must cite education and government, where, in general, productivity seems at best to have stood still or decreased over the past several decades. This does not mean that we will be unable to learn how to increase productivity in these service areas. It means merely that our present modes of organization are not such as to emphasize productivity or to take advantage of technology and capital, which of course have been the principal multipliers in the agricultural and industrial sectors. There is a second and very important effect. The growth is primarily in the non-profit sector of society.

In 1930, of the 44 million in the total work force, only about 12.5% of all workers were in the non-profit sector. In 1960, in addition to those in government, there were about 5.25 million other workers in the non-government, non-profit sector.[85] I have no later figures for the non-profit sector, but considering the areas that one knows have grown sharply, such as health, it is probable that the overall growth has been at about the same rate as in government. If so, the total number of workers in the non-profit sector were about 22 million in 1972.

This would mean that about 27% of the total work force in 1972 was in the non-profit sector. As I shall argue later, the profit mechanism is one of the most vital in assuring the generation of the productivity culture in our mixed economy, and the implications of this shift are very significant from the standpoint of effective mechanisms for continuing gains in productivity.

I am even more concerned, however, about the philosophical implications. Because they misunderstand the role of profits in our kind of society, it is my observation that a large proportion of those with whom I come in contact from the non-profit sector not only disdain profits as a mechanism but often are convinced that, in some fashion, profits represent an illegitimate gain on the part of business, a gain at their expense as consumers. They appear to believe they would obtain the goods and services represented at far lower costs if only the selfish and grasping profiteers were not involved.

The very efficiency of the profit-making sector is in fact making the shift to services possible and, in turn, decreasing con-

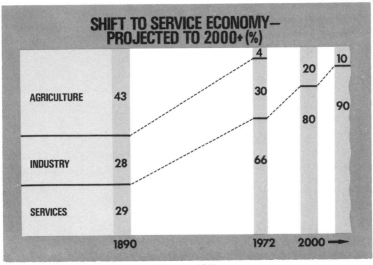

Figure 32[86]

stantly the number of men and women engaged in the production of food, goods, and that portion of the services sector
which is private enterprise with the consequence that they
become an even smaller portion of our total population.
Democracy being what it is, they also represent a decreasing
influence.

Further, because the society has wanted more education, it
has provided more educators and an undue percentage of them
display disdain and negativism toward the entire profit system.

If one concurs with John Maynard Keynes on the controlling
force of ideas in the long run, then these mistaken ideas, expressed often most articulately by an increasing proportion of
intelligent and competent individuals in the non-profit sector,
could portend serious consequences for the mixed economies.

The shift from agriculture and manufacturing to the services
sector is going to continue (Figure 32). High and increasing
productivity per person in the agricultural and industrial sectors
probably will bring the total percentage of the work force in
these two areas down to 20% by the end of the century and
perhaps to 10% or below not too long thereafter. That means
that 90% of our total work force will be in what we now describe
as the services area in only a few decades. If the shift continues
to take place, as it has, into service sectors with little or no

ninety-one

regular increase in productivity per person, the overall shift will be slowed down and so will our ability to improve our overall standard of living.

I would be particularly concerned if we continue to choose to have any considerable portion of these increasing services provided directly by the government because of the enormous increase in the percentage of the work force employed directly by government which would follow.

It is in light of the very real possibility, if we continue present trends, that within a few decades more than 50% of our work force will be employed directly by national, state and local government that I believe Samuelson's dismissal of Hayek's fears may be premature.*

*See page 66.

The Productive Society

III. Preserving The Productive Society

III. Preserving The Productive Society

In the preceding discussion, I hope I convinced you that there is a very strong correlation between a high-productivity society and real quality of life and that in the United States, or the other key mixed economies, neither finite natural resources nor overwhelming population growth is likely in any inevitable sense to limit our ability to improve real economic welfare over the foreseeable future. Further, the real needs of the three-quarters of the world's population that lives in the developing countries are so great that they have no choice except to opt for methods of political and economic organization that seem to promise them the most certain course to higher productivity and more affluence. Nor are the developed countries free from this pressure, since even here in the United States, as recently as 1971, nearly one-fifth of our families earned less than $5,000 per year, and about one-half under $10,000 per year.

Although the mixed economies seem to have met far better than any others these needs for improving productivity, the changes in the structure of our society that are steadily bringing an increasing proportion of the total work force into government payrolls, together with the seeming inability of the mixed economies to solve the problem of inflation, are threatening the structure of our society and even the likelihood of our remaining free.

In the discussion that follows, I will suggest a series of private, institutional, and government actions and approaches designed to help us follow that intricate path to a permanently productive society without its becoming a road to serfdom.

Let me start with a few fundamental principles.

1. We must so write our laws and organize our institutions as to achieve the maximum possible alignment, first, between the interests of each individual and his employer; second, between the interests of institutions which provide the overwhelming proportion of such employment and the State; and third, between the interests of the individual and the State.

2. In attaining this alignment, we should bias our laws and organizational design and control mechanisms (government or private) in the directions that emphasize individual initiative and responsibility.

3. We must so organize our society that the laws and regulations we write and the modes of organization we establish emphasize the individual citizen, first as a consumer, and second as a producer, and with care to observe the next two principles.

4. We must depend upon the market economy to the absolute maximum extent possible for goods and services.

5. Government interventions should be structured to provide overall rules within which the market economy can function most effectively. This includes the need for order in the society, stability in the currency, and mechanisms that assure both the right and responsibility of contract. Government interventions in the price structure should be used, but with care to cause it to reflect the value of such public goods as the environment, or where necessary for the welfare of society, in such matters as health and safety.

I will elaborate on these five principles as seems appropriate, although I have not done so in any order or sense of priority. For example, as an illustration relating to Principle 3, difficulties often arise in our kind of society because producers band together to get rules and modes of operation that favor them as producers at the expense of the individual consumer. Ordinarily, we consume as broadly diffused individuals. As such we do not represent an organized constituency in the political process. On the other hand, producers when banded together present a very strong constituency. They may be businessmen from some industry such as textiles seeking quotas, or farm organizations asking for guaranteed market at a guaranteed price from the government, or labor unions advocating laws to protect their jobs or assure a wage structure or work rules to their advantage.

As I heard Milton Friedman observe trenchantly a few weeks ago: "Actions like these that serve primarily the interests of those seeking them shouldn't surprise anybody, but it is up to the rest of us to see that they don't get their way."

Price controls may seem to serve well the citizen as consumer, but in fact they inevitably distort the supply/demand relationship eventually to the point where he is injured as a consumer and may be put out of work as a worker if he is in one of the sectors affected. Rationing gasoline in such a manner as to ensure that the average driver has enough to meet all his needs might seem an appropriate way to fulfill Principle 3 in this time of petroleum shortages. Since this would be quite likely to mean even fewer petroleum products for agriculture and industry, however, the end result would be just the opposite by increasing unemployment sharply.

There is every indication that the best overall balance among such conflicting interests is achieved where goods or services are supplied by private, profit-seeking organizations operating under a framework of law that provides stable government, prevents monopoly, protects the right of parties to enter into mutual contracts and the right to require performance in accordance with the terms of the contract. The general atmosphere should be biased toward success instead of failure. This kind of atmosphere exists in a growing mixed economy with proper dependence upon the market.

I think we have that kind of system now, but we may lose it if trends, which I described in Chapter II, continue to the extent that more than 90% of our workers are in the service sector and more than half are working for the government or non-profit institutions.

There is no real need for the trend to continue. Many of the present services performed by government or non-profits could be performed by profit-seeking industry. The same is true for a variety of the new services we are likely to demand. Hospitals need not be non-profit. The U. S. Postal System could be a private, regulated utility; so could the Federal Commercial Aviation services. Much of the total educational system could be in the profit-making sector. Public utilities rarely should be government owned.

There are two principal objections advanced against putting institutions such as these into the private, profit-making sector. The first is that profits should not be made on services such as schools or hospitals. Usually, this is related to the likelihood that important but costly services will be denied society because profit-making institutions will seek to maximize their margins, go after that portion of the market on which they can make money, and avoid that portion for which the services are so difficult or sufficiently rare that only losses are in prospect.

I suggest that most of these exceptions can be covered by a combination of approaches:

A. Fixed-price or cost-plus contracts can be entered between an appropriate government entity and such private profit-making institutions requiring that these services be performed. The contracts could then be administered under provisions common among federal agencies.

B. Some of the services could be required by regulation; i.e., for the privilege of serving a geographic area such as a state or region, services must be provided in all specified portions of the state or area at agreed upon or regulated prices.

C. The exceptions for which neither of these nor alternatives appear suitable could be handled by government-operated facilities.

The second general objection is that the profits will increase the costs of the service unduly. This simply reflects a complete failure to understand the role of profits in the market system. Lack of understanding is by no means confined to the non-profit sector in this country. I suppose the usual answer expected from the businessman when asked why he is in business is that he is there to make money—to operate at a profit. It may surprise you after my emphasis on the importance of profits to the system that I now assert as strongly as I know how that, however important profits are to the success of the institution, they are not its reason for existence.

Just so I may be as specific as possible, I will illustrate the principles I believe do hold as we have been applying them at Texas Instruments:

1. TEXAS INSTRUMENTS *EXISTS* TO CREATE, MAKE AND MARKET USEFUL PRODUCTS AND SERVICES TO SATISFY THE NEEDS OF ITS CUSTOMERS THROUGHOUT THE WORLD.

2. THE *OPPORTUNITY* TO MAKE A PROFIT IS TI'S *INCENTIVE* TO CREATE, MAKE AND MARKET USEFUL PRODUCTS AND SERVICES.

We do not exist to make a profit, but the *opportunity* to operate at a profit is the *incentive* to create, make and market useful products and services for our society. Furthermore, if TI is to prosper, to grow, to broaden, and to improve the products and services we offer, it must be so managed as to ensure a high level of long-term profitability. For the privilege of the incentive, in the form of profits, to create, make and market useful products, our economic system in the United States requires

ninety-seven

that we operate at a profit or disappear. The requirement that we make sufficient profit to attract the necessary investment or cease to exist provides the discipline we need to ensure that we operate effectively.

Certainly, at Texas Instruments the profit system automatically forces us to recognize increasing costs, either by increasing prices or improving productivity or both. Competition on a worldwide basis limits severely our ability to increase prices. In fact, in sectors of our business totaling something more than half our annual volume, we have had average price decreases over the past two decades of 15% per year. Consequently, we exert a constant effort to improve productivity, and we have been forced to learn how to use science and technology, capital and management to improve our effectiveness and reduce costs. On the record, there can really be little question that competition and the profit mechanism have provided a crude but effective servo-system which requires the private enterprise sector to respond to the pressures by steadily increasing productivity per person.

Every long-lasting institution evolves a culture of its own, and Texas Instruments is no exception. Our culture is determined by our policies, procedures, and practices both as they are formally stated and installed as systems and as they are actually perceived and executed by the individuals who make up Texas Instruments. These policies, procedures, and practices—and the culture they produce—all are aimed at creating, making, and marketing products and services to satisfy the needs of our customers around the world and are keyed to the incentive provided by the profits we make.

The relatively automatic operation of the market economy creates a culture common to all private enterprise, a culture that is dependent upon and oriented toward the need to provide products and services for customers at a profit. It is a culture in the full sense of the word, one that automatically biases the entire sector toward high and increasing productivities per person; and it does not exist in the not-for-profit sector.

In fact, profits do not cost the American people. They create savings for them. It is the bias toward constantly increasing productivities per person that has made possible the American standard of living. Further, profits have been like a catalyst in a chemical reaction. They have played a very important part in making the whole system work and in creating the gains in productivity. Yet, in themselves, they are a very small part of the price at which the product or service is sold.

Few realize just how small a part they are. Figure 33 shows the record for 1972, a relatively good year so far as corporate profits are concerned:

U.S. CORPORATE SALES & PROFITS, 1972

	BILLIONS OF $	PERCENT OF CORPORATE SALES
CORPORATE SALES	$1803	100.0
PROFITS BEFORE TAXES	98	5.4
Income Tax Liability	43	2.4
PROFITS AFTER TAXES	55	3.0
Dividends Paid	26	1.4
Estimated Income Taxes on Dividends Paid by Individuals	6	0.3
UNDISTRIBUTED PROFITS	29	1.6

Figure 33[87]

Corporate sales amounted to $1803 billion in 1972. The profits before taxes on those sales were $98 billion or only 5.4% of total corporate sales. The tax liability for corporations on those profits was $43 billion, leaving profits after taxes of $55 billion. The corporations paid out $26 billion in dividends, and I estimate that the taxes paid by the individuals receiving those dividends were $6 billion, so the recipients retained $20 billion and the corporation retained undistributed profits of $29 billion.

Thus, this magnificent, highly productive machine was kept running by only 5.4% before taxes of the selling price of all the products and services produced and sold. This is far from the end of the story, as Figure 34 demonstrates.

After paying income taxes, the corporations retained $29 billion. After paying the taxes on the dividends they received, the shareholders retained $20 billion. Thus, of the total profits

ninety-nine

DISTRIBUTION OF CORPORATE PROFITS, 1972

	BILLIONS OF $	PERCENT OF PROFITS BEFORE TAXES
PROFITS BEFORE TAXES	$98	100.0
Profits Retained by Corporations	29	29.6
Dividends Retained by Individuals	20	20.4
SUBTOTAL	49	50.0
Taxes Paid by Corporations on Income	43	43.9
Taxes Paid by Individuals on Dividends	6	6.1
SUBTOTAL	49	50.0

Figure 34[88]

generated, $49 billion were retained by the corporation or its shareholders.

On the other hand, the citizens of the United States, via their Federal Government and the states in which they lived, shared in those same profits in the amount of $43 billion paid by the corporation and an estimated $6 billion more paid by the recipients, for a total of $49 billion. The tax paid by dividend recipients is this low because a portion of the dividends went to other corporations, and the tax on this portion is included in the corporation tax amount. Thus, in fact, without contributing in any way to the $3 trillion of assets necessary for the corporations to function, the citizens of this country shared in the profits just as though they owned 50% of the shares of those corporations.

There is no need to talk about government ownership of corporations when the tax system functions in this fashion. Quite frankly, and primarily for this reason, I approve of the manner in which this tax function is performed, although I do believe the corporate taxes are somewhat higher than optimum and that this has contributed to the capacity shortages present in so many of our industries now. The profits earned a few years ago were simply not sufficient to justify added investment at the rate which would have been required to have capacity in place today. Unfortunately, the effect on supply and costs is compounded, because the new capacity inevitably takes advantage of the latest technology and usually produces more effectively than that in place. In the overall, the system would operate more effectively and generate higher productivities if the distribution to the corporation and its shareholders were higher, such as 60-40 rather than 50-50.

This illustrates the first principle I suggested. The profit and tax systems combined align the interests of individual shareholders, the corporation, and society as a whole, since all share in the success of the organization and since all are better off when it succeeds.

Now you would think that with the facts as they are, there would be intense interest in the profitability of the corporations that produce our goods and services and cheering in the streets when the profits are high, because increased corporation taxes reduce individual tax burdens. Unfortunately, the average citizen doesn't understand either the role of profits in making the system work or his own high share in them. In 1972, the typical American adult, responding to a national survey, stated that he believed the major United States corporation makes an after-tax profit of 28¢ on each dollar of sales.[89] Since the right number in 1972, a very good year, was 2.8¢, it is very clear that we in business are not communicating well with the general public!

If corporate profits were indeed 28% after taxes, some of the general public attitudes about the ability of corporations to pay higher wages, higher taxes, and absorb environmental costs, all at lower prices, make much more sense. Then after-tax profits in 1972 would have been $550 billion, or about half the gross national product, instead of $55 billion, and it might even have been possible to cure those shortcomings of our society, which in any sense spending money can cure, by soaking the corporation.

Max Ways in an excellent article in the September 1972 issue of *Fortune*, summarizes it this way:

"Much of the ill repute of business arises from. . . . misconceptions, which, in a complex, prospering society, are bound to occur unless there is a huge and intelligent effort to keep the people informed. The absence of such an effort explains why the reputation of business in the last twenty-five years has been getting worse, while the performance of business, by almost one hundred one any economic or social or moral standard, has been getting better."[90]

Edward B. Rust, president of the U.S. Chamber of Commerce, stated in a recent speech that corporate executives should be allies, not enemies, of consumer advocates, including Ralph Nader. Mr. Rust was agreeing with the third principle enunciated. He was emphasizing that the purpose of all businesses is found in the products and services they provide to their customers and that there can be no quarrel with consumer advocates who insist that the quality of the product or service provided meet the standards claimed by its supplier and those set by applicable law.

It is especially tragic when poor relationships between the consumer and the private enterprise sector that serves him result in unwise regulations and laws, adding to the present tendency to overcontrol and so threatening to destroy the system itself—tragic because no other system has served the consumer so well. Certainly, there is no indication that the centralized economies are likely in any way to be especially sensitive to the needs of the consumer. One small illustration appears in Dan Greenberg's *Science and Government Report*:

Ralph, They Need You in Moscow

Roadtesting of nine brand new Soviet-built Moskvich passenger cars has been abruptly cancelled by Britain's Consumers' Association following the discovery of numerous safety defects that caused the CA to conclude that the cars were unsafe even for its experienced test drivers.

'The Moskvichs, vanguard of a Soviet effort to earn foreign exchange in western consumer markets, were found to have leaking brake fluid, oil leaks onto brake linings, loose steering parts, loose wheel nuts, and other defects. The director of the Association said, 'We have found the potentially dangerous standard of construction of this car to be by far the gravest safety problem we have ever had with any car.'

'Noting that 20,000 Moskvichs have been sold in Britain, the Association demanded that the government order them off the road. To prove its point, it announced that it would bestow one of its Moskvichs on a government test facility.

'For safety purposes, transportation to the test center was by trailer.[91]

Elisha Gray II, recently retired chairman of Whirlpool Corporation, has long been an articulate advocate of an enlightened consumerism. Since his retirement, he has been responsible, as chairman of the board of the Council of Better Business Bureaus, for a major effort supported broadly by business and industry to strengthen the 140 Better Business Bureaus scattered in cities across the entire United States so they can deal more effectively with complaints and inquiries from consumers. This is an important effort because it concentrates directly on getting the consumer and his supplier together to solve whatever problem exists.

U. S. News and World Report, in its October 8, 1973, issue, reports on a considerable variety of similar significant efforts *"to raise the level of public confidence and erase widespread misconceptions about prices, profits, and performance."*[92] All described are useful and needed. Their impact will, however, be limited and, like these discussions of mine, probably will be viewed as advocacy initiated as a consequence of strong self-interest.

I believe that John Maynard Keynes' concluding sentence in his book, *The General Theory of Employment, Interest and Money*, I quoted earlier is right: "But, soon or late, it is ideas, not vested interest, which are dangerous for good or evil . . ."

Therefore, I suggest that what we need is the kind of comprehensive, analytical collection of coordinated principles with respect to the mixed economy that Adam Smith's *Wealth of Nations* provided for the laissez-faire economy. I do not want to appear to be denigrating the existent and often competing works of Friedman, von Mises, Röpke and Samuelson, among others. Perhaps I am simply ignorant or merely proving I am among those who, as Keynes said, "after 25 or 30 years of age are not influenced by new ideas." But I do not believe, brilliant as some of these works are, that any of them satisfy the need for a comprehensive statement that illuminates the whole of the philosophy, theory, technology, principles, and practices essential if the mixed economy is not just to survive but to evolve into the kind of productive society the requisites for which I am groping in this discussion.

I suppose there will be many who will object to my call for another *Wealth of Nations* because "it's obsolete," or "most of

one hundred three

what Smith thought he proved has since been disproved." In that sense, Keynes' *General Theory* or, for that matter, Darwin's *Origin of the Species* also are obsolete.

Perhaps I can make clear what I believe is needed by quoting from Max Lerner's Introduction to the Modern Library Edition of the *Wealth of Nations*:

That is why the arguments of all the scholars who have been thrashing about, seeking to determine how original Adam Smith was, are essentially futile. No first-rate mind whose ideas sum up an age and influence masses and movements to come is in any purist sense original. The Wealth of Nations *is undoubtedly the foundation-work of modern economic thought. Yet you can pick it to pieces, and find that there is nothing in it that might not have been found somewhere in the literature before, and nothing that comes out of it that has not to a great degree been punctured by the literature that followed. What counts is, of course, not whether particular doctrines were once shiny new, or have since stood the ravages of time. What counts is the work as a whole—its scope, conception and execution, the spirit that animates it and the place it has had in history.*[93]

It is a sense of the whole which must be comprehended, portrayed, and communicated so that the complex of laws, regulations, institutions, and procedures and practices necessary to the sound development of the modern, mixed economy may be orchestrated into an effective concert that avoids the discordant and self-destructive cacophony which is threatening to emerge.

A university should provide the best atmosphere for scholarly review and dispassionate judgment, plus advocacy of action flowing from that judgment, and I would hope that a modern-day Adam Smith is now at work in one of our universities preparing a treatise conveying this "sense of the whole" for the mixed economy. In its absence, I will suggest at a much more modest level some directions to follow and actions to pursue which would, I believe, enhance the performance of our own society.

Edward Denison has made extensive studies of the sources of economic growth in the United States and other countries. His findings from one such study are summarized in Figure 35.

United States: Sources of Growth of Total National Income and National Income Per Person Employed, 1950-62

Sources of Growth	Contributions to Growth Rate in Percentage Points	
	Total National Income	National Income Per Person Employed
NATIONAL INCOME	3.32	2.15
Employment	.90	—
Hours of Work	- .17	- .17
Age – Sex Composition	- .10	- .10
Education	.49	.49
Total Labor Input	1.12	.22
Total Capital Input	.83	.60
Total Land Input	.00	.03
Total Factor Input	1.95	.79
Advances of Knowledge	.76	.75
Improved Allocation of Resources	.29	.29
Economies of Scale	.36	.36
Irregularities in Pressure of Demand	- .04	- .04
Output Per Unit of Input	1.37	1.36

Figure 35[94]

Over the period 1950-62, national income in the United States grew at 3.32%. All input factors contributed 1.95% and increases in output per unit of input contributed 1.37%. Similarly, national income per person employed grew 2.15% over this span of years. All input factors produced .79%, and increases in output per unit of input, 1.36%.

There are many interesting implications implicit in the distribution of these sources of growth, but I especially want to emphasize two as illustrated in Figure 36.

Note that education contributed 0.49% and advances of knowledge contributed 0.76%. Over in the national income per person category, 0.49% came from education and 0.75% from advances of knowledge. Thus, about 38% of our increased growth in national income and 58% of the increase in national income per person employed came through aspects of learning via the impact of education on our labor force and applications of knowledge such as technological innovation and improved management.

Denison is not alone in his conclusions. Robert Solow, after analysis, concludes that: *"Scarcely half of the increase in America's productivity per capita and in real wages can be accounted for by the increase in capital itself. More than half of the*

one hundred five

United States: Selected Sources of Growth of Total National Income and National Income Per Person Employed, 1950-62

Sources of Growth	Total National Income		National Income Per Person Employed	
	Contributions to Growth Rate		Contributions to Growth Rate	
	In Percentage Points	Percent of Total Contribution	In Percentage Points	Percent of Total Contributions
NATIONAL INCOME	**3.32**	**100.0**	**2.15**	**100.0**
Education	.49	14.8	.49	22.8
Advances of Knowledge	.76	22.9	.75	34.9
Total Educational Input	1.25	37.7	1.24	57.7
All Other Sources	2.07	62.3	.91	42.3

Figure 36[95]

increase in productivity is a 'residual' that seems to be attributable to technical change—to scientific and engineering advance, to industrial improvements, and to 'know-how' of management methods and educational training of labor.''[96]

Denison also studied the sources of growth in each of seven countries of Northwest Europe over the same time period using the same methodology, (Figure 37).

These countries of Northwest Europe—Belgium, Denmark, France, Germany, Netherlands, Norway and the United Kingdom—as a group had much higher growth rates during these years than the United States both in national income (4.76% vs 3.32%) and in national income per person employed (3.80% vs 2.15%) and Denison was attempting to identify the sources of growth that contributed to this increased rate. Advances of knowledge played about the same part, 0.76%. Education was less important at 0.23%. Very large gains, accounting for the higher growth rates, appear to come primarily because the economies of these countries were not nearly so well developed, and there was much more room to improve them. For example, in national income 0.56% came from changes in the lag in the application of knowledge, general efficiency, and errors and omissions; 0.68% came from improved allocation of

Northwest Europe*: Sources of Growth of Total National Income and National Income Per Person Employed, 1950-62

Sources of Growth	Contributions to Growth Rate in Percentage Points	
	Total National Income	National Income Per Person Employed
NATIONAL INCOME	4.76	3.80
Employment	.71	—
Hours of Work	- .14	- .14
Age – Sex Composition	.03	.03
Education	.23	.23
Total Labor Input	.83	.12
Total Capital Input	.86	.65
Total Land Input	.00	- .04
Total Factor Input	1.69	.73
Advances of Knowledge	.76	.76
Changes in the Lag in the Application of Knowledge, General Efficiency and Errors and Omissions	.56	.56
Improved Allocation of Resources	.68	.68
Balancing of the Capital Stock	.08	.08
Economies of Scale	.93	.93
Deflation Procedures and Irregularities in Pressure of Demand	- .06	- .06
Output Per Unit of Input	3.07	3.07

*Belgium, Denmark, France, Germany, Netherlands, Norway & United Kingdom

Figure 37[97]

resources; and 0.93% from economies of scale.

This is a very important difference because, from it, I would draw the conclusion that the more advanced and the more mature an economy, the more significant the part played by education and advances in knowledge as sources of growth. These findings corroborate the experience and intuition of those of us in technologically based industry and also suggest areas where government intervention can make a major contribution to growth.

The change in average educational level in our society (from grade school to high school graduate in 42 years) accompanied and contributed to our increase in affluence. For the individual, education has surely brought great rewards beyond the improvement of material life. The opening of the mind, inciting and satisfying curiosity, broadening knowledge, are important contributions to his standard of living. For the employing institutions, an intelligent and educated labor force can be a key factor for satisfactory performance and growth. For society and the state, an educated citizenry is a necessity for the preservation and improvement of the Republic. The overall impact of mass education is one of the great success stories of this country's development.

one hundred seven

Yet, there are problems in our system of education. We may have become so intoxicated by the success story that we have continued to follow the pattern that produced the success to a point of sharply diminishing or even negative returns. Earlier, I pointed out that 27% of our total work force in 1972 was in the non-profit sector. A very large proportion is in education. (See Figure 38.)

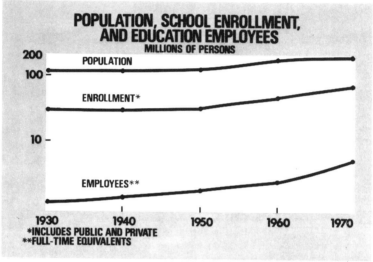

Figure 38[98]

Between 1930 and 1970 our population grew from 123 million to 205 million—about 60%. During that same span of time, our school enrollment didn't quite double, advancing from 29.7 million to 58.7 million. But the number of full-time equivalent men and women employed in education grew more than four times, from 1.3 million in 1930 to 5.4 million in 1970. Thus, in the last 40 years, the number of full-time employees in education has grown twice as fast as the number of students and more than three times as fast as the population.

Expressed in 1970 dollars, while school enrollment about doubled between 1930 and 1970, our total expenditures for education at all levels grew more than nine times, from $7.5 billion to $70.6 billion, and our expenditures per student grew nearly five times, from $253 per student in 1930 to $1203 in 1970 (Figure 39). The biggest increases have come since 1950 and we

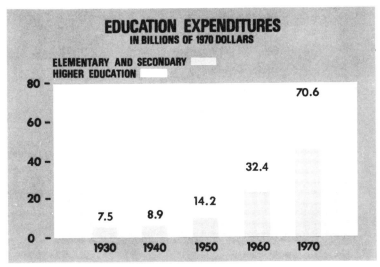

Figure 39[99]

now spend nearly three times as much per student in constant dollars as we did in 1950.

These comparisons are not completely fair because the number of students in higher education, where the costs per student are obviously much higher, has increased more rapidly than in elementary and secondary schools. More detailed examination will show, however, that the increases in costs for elementary and secondary education were only moderately lower than those in higher education.

As to the nature of the forces generating these disparities, let me quote from "The Economics of the Major Private Universities," a paper by Dr. William G. Bowen, Provost of Princeton University and Professor of Economics and Public Affairs:

Let us imagine an economy divided into two sectors, one in which productivity is rising and another in which it is constant, the first producing automobiles, and the second, 'education' (defined as some amalgam of students and knowledge). Let us suppose that in automobile production output per man-hour increases at an annual rate of 4 percent, compared with a zero rate of increase in the education industry. Now, let us assume that money wages in the automobile industry go up at the same rate as productivity in that industry. This means that each year

one hundred nine

the typical auto worker's wage goes up by 4 percent, but since his output increases by exactly the same percentage, the labor cost of manufacturing a car will be unchanged. This process can continue indefinitely, with auto workers earning more and more each year, with costs per car remaining stationary, and with no rise in automobile prices necessary to maintain company profits.

'But what about the education industry? How it fares in this imaginary economy depends on what assumption is made about the relationship between increases in faculty salaries (treated, for the sake of simplicity, as an index of all salaries in the education industry) and the increases in the wages of auto workers. Over the long run, it is probably most reasonable to assume that faculty salaries increase at approximately the same rate as wages in other sectors . . . .

'If the salary of the typical faculty members does increase at an annual rate of 4 percent, so that his living standard improves along with the living standard of the auto worker, but if output per man-hour in the education industry remains constant, it follows that the labor cost per unit of educational output must also rise 4 percent per year. And there is nothing in the nature of the situation to prevent educational cost per unit of product from rising indefinitely at a compounded rate of this sort.

'The particular assumptions included in this analysis are, of course, merely illustrative, ... and the numerical results can be changed by assuming a different rate of productivity increase and a different rate of increase of money wages in the non-educational sector, by assuming that faculty salaries increase at a somewhat different rate from money wages in general (either faster or slower), and by allowing for some increase in productivity in the field of education. But modifications . . . will not alter the fundamental point of the argument, which is that in every industry in which increases in productivity come more slowly than in the economy as a whole, cost per unit of product must be expected to increase relative to costs in general. Any product of this kind—whether it be haircut, custom-prepared meal, a performance of a symphony concert, or the education of a graduate student—is bound to become ever more expensive relative to other things.[100]

The pressures Dr. Bowen outlines would apply if the ratio of students per employee in the education industry were constant. Since in fact our educational system consistently has decreased the number of students per employee in the system, a double compounding has been taking place.

It has been pretty much doctrine that all that was necessary to solve the problems of the education industry was sufficient money. It would seem to me that the events of the past 20 years should have dispelled that illusion completely. After all, spending approaching three times as much per student (in constant dollars) does not seem to have improved the effectiveness of the system or the quality of the average end product to anything like that extent.

More money will help, but not much, if the record of the past 20 years means anything! Remember Dr. Bowen's observations. If the productivity in the rest of the economy is increasing and that in education is not, then it will take a considerable amount of money per year merely to keep the pay scales of those in education comparable to those in the rest of society, even though nothing more is being produced for the additional funds.

There are four vital points which must be made:

(1) Any questions raised with respect to decreasing productivity per person within education itself are not about the advantages of education for all but only about the institutions and the procedures—with how and what and when we have chosen to provide the education. Education for all to the maximum of capability and desire is a proper objective in that higher standard of living we seek for each person.

(2) As Denison's studies emphasize, the constantly improving educational level of our total population contributes appreciably to the increasing productivity per person in the private sector. Thus, to the extent that the additional costs go to training a larger percentage of the population to a higher level of education, the education industry is entitled to a share of the productivity gains made by the private sector.

(3) The problems within the educational system which have produced the constantly decreasing productivity are at least as much the fault of society itself as they are of the education industry. After all, we outsiders make up most of the population, and we elect the school boards and set the general

standards and specifications of the system within which the teachers operate. It is that same greater society that is responsible for the turmoil, the problems of race and discrimination, the imposed solutions such as busing, three-month summer vacations, and a variety of other limitations, strictures or frictions which both create the system within which education operates and interfere with its operating effectively.

(4) This unhappy state of affairs is not because the professionals in the education industry are less able or more venal than the rest of us. On the contrary, I suspect that the average professional in education works as hard, is at least as well trained, and if anything, more dedicated than those of us who operate in the areas where productivity has been improving.

There can be no escape from the constantly escalating costs if the only solution is the expenditure of more and more man-hours of adult instructional and administrative time per student hour. Somehow, we must find ways to maintain and improve the quality of education with fewer adult hours per student hour.

Most of the discussion thus far has been discouraging. Indeed, Dr. Bowen, in the remainder of the paper referenced, concludes that educational costs will go on increasing relative to the rest of society, simply because productivity can't be improved at an adequate rate.

Frankly, I cannot agree with that conclusion. It is inconceivable to me that if we really want to, if we apply the multitude of talents we possess as a society, and if all who are in the profession of education apply their skills enthusiastically, we cannot get 2% to 5% more work done each year than we did the year before. That is all it takes—3% more productivity per person per year would keep up with the rest of society; 5% more productivity per person per year would generate a lead over the remainder of society and produce some surplus funds which, in turn, could be used to improve the quality of education itself without increasing real costs.

Before anything else, we need to ask ourselves if the kind of education we are providing now is right for all of the more than 15 million teenagers who presently are in high school.

According to the U. S. Office of Education, 60% of this year's high school graduates will enter college. In another 10 years, the figure probably will be nearly 70%. Now we have about 9 million in colleges and universities; assuming present trends, in another 10 years we'll have more than 12 million. Is the only route to a college-level education to continue to coop up 12 million of our young people in the classrooms and relatively artificial atmosphere of school for so much of their lives?

Mightn't it be wiser if, beginning at 15 or 16, those who wished could go to work for four to six hours a day and attend classes with fellow workers for two or more hours, probably via television classrooms near their work places, or do so alone at their work place or at home, using TV and audio cassettes plus supplemental material?

In the North Texas region we have a closed circuit TV network with talkback capabilities coupling nine universities and colleges and 46 classrooms in seven industrial organizations in a complex of nine locations around Dallas, Fort Worth and Sherman, Texas. The TAGER TV network opened in September 1967 and presently is in its sixth year of operation. In the fall of 1973, a total of 71 graduate and undergraduate-level courses in business, engineering, science, and mathematics were offered to a course enrollment of approximately 1600. At Texas Instruments alone, 317 were enrolled in programs, most working toward master's degrees. This is out of a total TI professional population in the Dallas area of about 6000, of whom about 1900 already have masters or doctorates. The program is an overwhelming success; yet, we have not designed an educational system based on TV. We really use TV just to expand the reach of every participating educational institution to include the classroom in our plant. Although there is some influence on the graduate-level subject matter by the industrial organizations, in general, the schools are establishing the curricula and the companies are furnishing the classrooms, the students, and most of the funds.

Further, for all my emphasis on the necessity for considering cost and although costs were considered in establishing the network, it was not designed with specific and reduced costs per student hour in mind. Actually, the cost per student instructional hour is about comparable with normal university costs. However, the convenience of the classroom locations and the scheduling of classes throughout the working day and evening ensure that many more workers in the participating companies go on to post-graduate degrees and that the cost to the companies in lost working hours is markedly reduced.

In addition to the television-based educational program conducted with local universities described, Texas Instruments operates a Learning Center providing training and job-related knowledge and skills without high school or college credit as a consideration. The Learning Center is a multi-media facility. It uses video tape, audio tape, and computer-aided instruction supplemented by live lectures, class discussion, and the usual printed reference material. A balanced mix of media is maintained although video tape is primary. Practically all of our facilities, including sales offices, are equipped with video playback equipment, so most of the Learning Center program can be offered at TI facilities anywhere in the world. An extensive array of courses is offered in basic and advanced electronics, manufacturing and production systems, marketing, data processing, management and supervision, and in such general fields as improved reading skills, introduction to mathematics, statistics, accounting, zero-base budgeting, linear programming, and simulation. (See Figure 40.)

The spectrum of courses available in data processing, for example, is sufficient to enable individuals to qualify as operators, programmers, and functional or technical analysts.

During 1973, more than 1400 students completed about 25,000 student hours of instruction. About 10 percent of the students completed the courses by themselves, self-paced and self-motivated. We estimate that in 1974 more than 5700 students will complete over 100,000 student hours of instruction. Of these, about half will work on a self-paced basis.

On the basis of our training experience at Texas Instruments, there simply is no doubt that a carefully prepared instructional program on TV cassettes, plus the requisite supplementary written materials for reference and testing, with occasional tutorial help, is superior to conventional classroom teaching for much subject matter. It does have to be well done, and the student must want to learn; but those are requisites for any successful program of education.

I suggest that we need a radical revision in the institutional approaches we use to provide broader education to massive groups of people. I believe that our limited experience in Dallas,

Educational and Training Courses for TI Employees

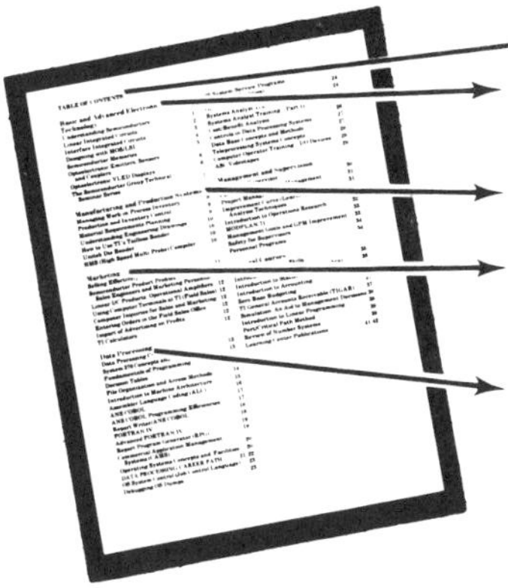

Figure 40

substituting TV in plants and offices for classrooms in universities, and using TV and audio cassettes, demonstrates that for many, if not most students, required in-school classroom attendance could be terminated after the tenth year. Thereafter, the education requirement for these students would be fulfilled in close connection with their jobs and at their place of work through highly flexible programs emphasizing the use of TV cassettes as well as TV classrooms operated as adjuncts to live classrooms. The program would be open-ended so that one could continue working through any requisite number of years to attain various diploma levels from high school up. The course content quite properly should be set both to expand cultural horizons and to augment the career being pursued. This should be accomplished with the collaboration of the organization at which the individual is working, and the diplomas would be granted by the appropriate degree-granting institutions.

Since the individuals concerned would be working in both business and non-business organizations, and since anyone who wanted to do so could proceed as far as his competence and desire led him, a large part of the status-based social compulsion to complete first high school and then college in order to "belong"—to be "eligible" for a suitable career—would be

one hundred fifteen

removed. Actually, most vocational courses appropriate to the work done probably could be given more effectively this way than the present system allows. This is at least as true for such college-level vocational training as cost accounting or tax law as it is for study we customarily think of as being at the trade school level; for example, machine shopwork or office procedures.

One thing that definitely bothers me about our present system is that, even in industry, the road to anywhere near the top from the shop is becoming more and more difficult to traverse. The present mechanisms ensure that such a large percentage of those likely to succeed in management will have gone to college that, in general, enough college graduates will be put in the very lowest kind of supervisory positions to acquire experience, and some of them will proceed from there to the top. If the kind of open-ended educational process described could be established, then presumably a fairly sizable number with adequate ability would start in shop jobs while comparatively young, acquire their college educations or equivalents along the way via a combination of experience and the organized but nonresidential kind of programs suggested, and once again there truly could be people who progress from the shop to the top —and be much better prepared in addition.

Indeed, as we all know, adequate preparation for a career in today's complex society is a process that needs to continue throughout one's working life. There is no conceivable way that packing all formal education into one's early life, terminating it with a diploma, whatever the level, in the Teens or early Twenties, and then going to work can provide the best career preparation and development. The new educational system described, where many fellow workers of varying age and experience are continuing their formal education in such a visible, convenient, part-of-the-working-environment way, would induce a similar interest in a larger proportion of all of us who work. A properly structured curriculum would ensure that, besides career-oriented studies, a suitable proportion of cultural and social courses would be available as well. Surely, this kind of education system in addition to enriching the individual would improve not only the productivity of the educational system itself but that of the entire society.

I would not suggest that the coordinated work-education program outlined should replace completely our present system, but I do believe that half or more of those attending high school and college would be better trained through this new system. For example, why couldn't at least half of the nation's supervisors, accountants, lawyers, engineers, salesmen, skilled craftsmen, secretaries (and for that matter, nurses, social workers and executives) get their education this way? Presumably, a large percentage of the training would be in response to specific need and thus help to balance the output of the educational system. Since it would be more focused and coupled closely to the individual's growing learning and interests—job connected and otherwise—it would surely provide for more effective learning and hence improve the productivity of the educational process that way too.

It is highly probable that, just as at TI, much of the cost of this educational program would be borne directly by the institutions employing the individuals because of the gains in productivity produced. To the extent that this is true and that the part of the education being paid for by industry replaces education presently being given in our schools, it will be at zero cost to society!

Milton Friedman, especially, has long advocated a change in the way we support educational institutions. He suggests that schooling be financed by giving parents vouchers, redeemable for a specified maximum sum per child per year if spent on approved educational services. Parents would be free to spend this sum, plus any additional amount they choose, to purchase educational services from approved institutions of their own choice. As Friedman envisioned the educational voucher program, the educational services could be rendered by non-profit institutions as they are now or by private enterprises operated for profit. This is very similar to the educational program the United States offered to veterans after World War II.[101]

There has been a great deal of debate already in the United States about educational vouchers. Proponents argue:
—The accountability of schools to parents and (at age levels where it is meaningful) to students will increase;
—Parents, having had to make a conscious act of choice, will be inclined to become more involved in other ways in the education of their children;
—The competitive nature of the market place and the lower "marginal cost" of obtaining private education will provide incentives for diversity in educational programs among both

one hundred seventeen

public and private providers, but also the more efficient provision of educational services measured by such criteria as increased consumer satisfaction and increased educational achievement for lower cost than otherwise; and
—The mechanisms of the marketplace will allow the educational system to respond more easily to outside social, cultural and economic stimuli and therefore to regenerate itself more effectively.

In opposition to these assertions, the critics of the concept (themselves as diverse in their values and political outlook as the proponents) have argued that the nation's educational system, with its predominant reliance on public providers, developed in response to a variety of social, cultural and economic needs over the Republic's history and should not be easily or lightly abandoned. In effect, they argued that education is a public matter in which, as in any other area which profoundly touches the commonweal, the allocation decisions should be made through the political process. As examples of this principle, the opponents argue that:

—Vouchers could easily result in greater racial/ethnic and class segregation by allowing parents to enforce more easily their preferences in such matters;
—Given that religious affiliated schooling is the main existing alternative to public schools, vouchers would lead to public subsidy of religious education, a condition which they argue is not only undesirable public policy but also quite probably unconstitutional;
—In competing for the voucher resources, some school administrators would be as unscrupulous in their claims as some vendors in other marketplaces and, thus, hucksterism would pervade the education system;
—It is at least questionable whether parents, especially low income parents, have the time, interest, or capability to make educational decisions for their children; and
—Private school participants either through tuition demands beyond the value of the publicly subsidized voucher or through admissions criteria, however neutral on their face, could exclude educationally or economically disadvantaged students; thus the public schools will become the "schools of last resort."

It is not my purpose here to argue either for or against educational vouchers. I do believe that experiments should be conducted to test the validity of the concept, however, because to the extent it can be implemented successfully, an educational voucher plan would satisfy the fundamental principles outlined for the preservation and enhancement of a productive society and decrease the pressure for constant growth in the nonprofit sector.

Very late in 1969, the Office of Economic Opportunity of HEW commissioned the Center for the Study of Public Policy to develop a model for testing the educational voucher system. Since then, in September 1972, the Alum Rock Union School District in California initiated an education voucher project, now under the sponsorship of the National Institute of Education, and confined to public schools. Obviously, it is far too early to draw any conclusions about educational vouchers, both because this single experimental site has been in operation such a short span of time and because it provides such a narrow base on which to make judgments. Apparently, however, on the basis of published reports, the school district itself, teachers, and parents are sufficiently enthusiastic to want to continue and enlarge the test. Certainly, the entire conception of experimental programs to develop information from which sound judgments can be made on these complex social problems is one that deserves encouragement.

It does appear that the educational voucher system might fit very well with the combined work-education program described beginning at perhaps age 15 and continuing through life. Vouchers might be used to cover costs to the same extent they are presently covered in government-supported high schools, colleges, and universities. Racial integration, for example, substantially disappears as a deterrent to the use of vouchers. Indeed, I would think that a coordinated work-education program of this kind would be of especially great value to families at the lower end of the income scale, because it encourages the teenager to begin to work and add to the family income. It does so while simultaneously opening up for those choosing to work a continuing educational opportunity instead of, as at present, cutting it off, except for the extraordinarily committed.

This kind of work-education program, carried through life, readily available everywhere, constantly broadening in scope as the system builds, combining as it could every kind of governmental and private institution and paid for by a combination of education voucher, employer, and individual support, would allow government interventions on any scale and at any point of the educational cycle. It would do so without forcing government ownership and government payrolls for almost the entire educational establishment as does the present system. Government, for example, could contribute to the cost of the development of the TV and audio cassette programs and to the development of TV links into areas which are sparsely industrialized where employer sites might not be large enough or so scattered as to make an economic program otherwise unworkable. Vouchers could be vehicles for retraining workers in industries phasing out or in cities where unemployment is high.

One of the immediate objections will be that we will be unable to find jobs for the millions of teenagers that this would throw into the labor market. Handled properly, I believe it would have just the opposite effect, but prior to a more detailed discussion on this point, it is necessary to return to the problem of inflation.

Referring again to Figure 27, there were marked increases in consumer prices for the countries listed over the past year; for example, 5.7% here in the United States; Japan, 11.9%; Finland, 12.1%; Spain, 12.1%; Greece, 13.1%; to Yugoslavia, almost 20%.

Further, the high inflation rates in the mixed economies seem to be associated with the laudable and indeed necessary stress on full employment—or really minimal unemployment. But what is minimal unemployment? In the United States there seems to be a strong correlation between excessive inflation and economic growth at levels which reduce unemployment rates below 4 percent.

Figure 41 shows that the annual increases in the consumer price index stayed around 1.0% to 1.5% during the early Sixties, climbed as the unemployment rate dropped to 4% and below, and reached 5.4% in 1969 when the unemployment rate averaged 3.5%. As the economy softened, in 1970, the unemployment rate climbed once again, reaching a high of 5.9% in 1971;

Figure 41[102]

and in October 1973 it was 4.5%, the Administration's goal for year-end. Unfortunately, this reduction in the rate of unemployment once again has been accompanied by inflation with consumer prices climbing 5.9%, up from 1972's low of 3.3%.

Yet, as Figure 42 portrays, unemployment rates of 3.5% to 5% are very high for industrial nations. The rates for other countries are not the ones usually published but are adjusted to U.S. concepts. Particularly striking are the very low percentages of unemployment in Japan and Germany. Admittedly, the special problems relating to the large non-white population in the United States are partially responsible for these differences; yet, even for whites only, our unemployment rate generally is the highest.

The dilemma occurs, of course, because in the mixed economies we use fiscal-monetary policy to try to regulate both inflation and unemployment; and since unemployment is perceived as creating innumerable social problems and being the greater evil, the stress in fiscal monetary policy is inevitably primarily on eliminating unemployment. The consequence is high inflation.

Elsewhere, I have recommended that this Administration initiate an extensive study program looking to comprehension and improvement of the employment and unemployment patterns

one hundred twenty-one

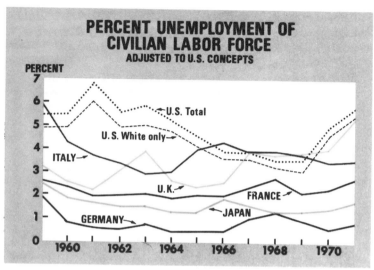

Figure 42[103]

in the United States and their meaning as related to other industrial nations. An example of the kind of study I have in mind is
one done by the Urban Institute for the U.S. Department of
Labor. The authors make some very significant observations.

Figure 43 shows the so-called Phillips curve relating inflation
and the unemployment rate. The relationship between price
inflation and unemployment over the years 1954 through 1970,
as plotted, seems erratic if the relationship is understood as a
static trade-off between inflation and unemployment. The authors, however, in their statistical study, relate these short-term
irregularities to recent inflation and unemployment history.
Thus, they point out:

. . . The economy experienced temporarily low inflation in
1955 relative to the long-run relation when unemployment fell
sharply because of continuing effects of the low inflation rate in
1954. Similarly, the temporarily high inflation in 1958 reflects a
carry-over effect from the high inflation rate in 1957. In 1956, with
only a small decrease in unemployment there was a big jump in
the inflation rate, reflecting in large part a delayed response to
the drop in unemployment a year earlier.

'When points are close to the long-term relation and there is
little change in unemployment, the inflation rate changes little.
See 1956 and 1957, 1959 and 1960, and 1962 and 1963.

'In years that were far from the long-run curve, big movements occurred in the inflation rate toward the long-run equilibrium even though there were small changes in unemployment. See 1955 and 1956, and 1966 through 1969.

'In the late sixties the country reaped the benefits of the costs that were incurred from 1958 through 1963. Starting from a long-run equilibrium combination of 5.7 percent unemployment and 1.7 percent per year inflation in 1963, expansionary fiscal-monetary policy reduced the unemployment rate to 3.8 percent in 1966 with a temporarily low inflation rate of 2.7 percent. This was the golden age of the New Economics. The rise in government expenditures without an increase in tax rates pushed the unemployment rate down to 3.5 percent in 1969 and the inflation rate rose to 4.7 percent. However, this was still below equilibrium on the long-run curve. If unemployment had remained constant at 3.5 percent, the inflation rate eventually would have risen to 5.9 percent per year.[105]

They observed that contrary to the classical economic view, there is a long-run relation between a money variable such as the rate of inflation and a real variable such as the rate of unemployment, because: "The real resources consumed by the

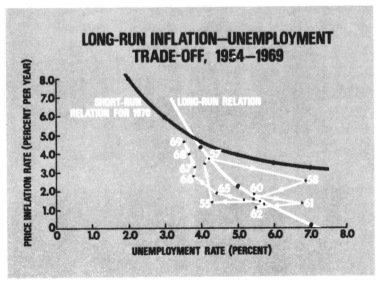

Figure 43[104]

frictions of the inflation process are sufficient to yield a sloping long-term Phillips relation."[106]

This is because

. . . *wages and prices typically are constant between discrete jumps and these jumps consume real resources. The higher the rate of inflation, the more frequently the jumps occur. Such wage and price changes inevitably trigger conflicts over the distribution of real income, and over market shares by firms and industries. These changes incur real costs in the form of market search, negotiations between employers and employees and between buyers and sellers, collective bargaining, strikes, lockouts, and job and location changes. The frictions caused by these jumps would normally slow price and wage movements, and thus inflation, if they were not accompanied and sustained by continuing excess demand.*[107]

They then conclude:

If the long-run trade-off that we have estimated is anywhere near correct, no manipulation of fiscal and monetary policy can resolve the dilemma of inflation and unemployment in the United States. Excursions out short-run curves will yield inflation-unemployment outcomes that deviate from the long-run relation, but only temporarily. However, structural policies such as manpower policy have the potential for reducing the long-run levels of both inflation and unemployment.[108]

These are extremely important observations, since the friction in the system consumes real resources without any increase in output, and if the friction could be decreased, the resources so released would not only decrease the inflationary pressures but simultaneously increase real output.

Friedman argues that "there is no perpetual trade-off between inflation and unemployment," and that "the trade-off is between acceleration of inflation and unemployment, which means that the real trade-off is between unemployment now and unemployment later." Thus the important policy is "to move to a sustainable rate of monetary growth and hold it there."[109]

Friedman's convincing arguments as to the importance of monetary policy have amended the economic views in this country so that now nearly all would agree that monetary policy is important but not necessarily to the point of saying, as Friedman does, that only monetary policy counts as far as inflation is concerned. It is difficult to deny the probable reality of the frictional consumption of resources in any real-world situation involving inflation and, if so, the unemployment rate arrived at by depending almost wholly upon monetary policy to achieve stability is likely to be too high to be politically acceptable.

The significance of this will become clearer if we examine the make-up of unemployment in this country in greater detail (Figure 44).

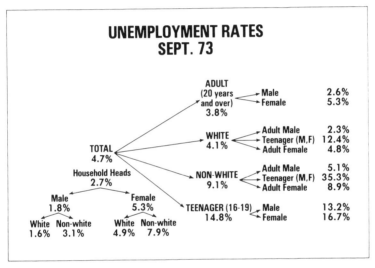

Figure 44[110]

Note that the unemployment among heads of households is only 2.7%, which for all practical purposes is no unemployment at all, and is accounted for primarily by short lags between job changes. This, of course, confirms the experience of anyone who has attempted to hire over the past year. Even for the non-white male head of household, the rate, although higher than for whites, is still only 3.1%. The real problems occur among the unemployed young and particularly among the non-white, unemployed young, where the rates are 14.8% and

one hundred twenty-five

35.3%, respectively. Further, the higher rates among adult non-whites are heavily related to lower levels of skill and education.

Can you imagine the long-term impact on this kind of structural unemployment of the lifetime work-education program suggested? There are a total of 1.21 million teenagers in this unemployed category of whom more than half are seeking only part-time work. If 605,000 of these young people were working, the unemployment rate for 16- through 19-year-olds would fall to 7%, a very tolerable number, and the overall rate would drop from 4.7% to 4.0%. If one assumes that, over these ages, one-half time was spent working and one-half time was spent in the accompanying educational program, then only half this number of jobs—or a few more than 300,000—would be required; or since we have a total employment of about 85 million, about a 0.3% increase in the number of jobs!

If at the same time the minimum wage rate was either eliminated completely or at least dropped considerably to reflect more nearly the low levels of skills these same young people can bring to the market place, their employment would be made economically attractive and also would help compensate for the inefficiencies introduced by the shorter working hours.

Since these teenagers aren't working now, even a rate of only $1.00 per hour—if they worked 1000 hours per year or fifty 20-hour weeks—would bring $605 million to their families. The benefits to the teenagers and their families actually would be well above the $605 million because of the indirect benefits such as hospitalization insurance, life insurance, profit sharing, and contributions toward pensions which would begin to accrue.

Furthermore, at a $1.00 per hour labor rate, some of the labor-intensive operations set up in Mexico and other countries to meet competition would be cost effective here in the United States.

With proper monetary fiscal policy, employment of these teenagers should have little or no impact on unemployment rates for adults. In addition, the difficulty of the problem will decrease in the future, simply because of the shift in age distributions which will take place. Presently 16/19-year-olds make up 7.7% of our total population. By 1985, it is estimated that 16/19-year-olds will make up only a shade more than 6% of our total population, and there probably will be 2 million fewer 16/19-year-olds than this year's 16.2 million.[111]

Of course, it is highly probable that if we had an attractive education/work program, there would be even more teenagers seeking to enter the program than are now listed in the unemployed category seeking work and this, in turn, would mean that a larger number of jobs for teenagers would have to be located to bring the unemployment rate down to the levels calculated. I couldn't view this with anything but satisfaction, however, because it would represent a clear indication of the success of the education/work efforts; and since the program inevitably would take a long time to implement, there also would be a long time to develop the jobs necessary.

Interestingly enough, *Science Magazine*, in its October 12, 1973 issue, includes discussion on the report, "Youth — Transition to Adulthood." This was the report of a panel of the President's Science Advisory Committee chaired by Dr. James Coleman. The review points out that the study was not to be on the problems of youth, but quoting Dr. Coleman:

. . . . *on the institutions that handle young people in our society, the institutions through which our young people reach adulthood. The object was to raise the question of appropriateness of the institutional experience young people have in becoming adults. . . .*

'*The main idea,' says Coleman, 'is that we have moved rapidly from a period when young people went to work fairly quickly after they became physically able to work. Now they are held out of (productive work) in special institutions. These special institutions are schools, and young people have the special role of students. We do not think this special role prepares them for being adults.*

*   *   *

In general, the recommendations are that young people have the opportunity for more nonacademic experience, for more contact with other age groups, and for more scope to make decisions for themselves and take responsibility for others. The panel would like to see the development of more specialized schools where students would follow particular interests, and of smaller schools to mitigate the impersonality of the prevalent big, comprehensive high schools. The panel would also like to

one hundred twenty-seven

see young people take roles other than as students, for example, as tutors of younger children.

'The panel sees schools acting as agents for young people not only in arranging work experience in conventional jobs but also, for example, in cultural institutions such as museums. Variations in the pattern of education, with stress on work-study programs, are regarded as particularly important.

'The report urges serious reexamination of laws which now protect workers under 18. It appears that some of these laws reduce opportunity for youth. It is suggested, for example, that there might be a dual minimum wage, with young people receiving a lower wage than adults, since a high minimum wage is regarded as a disincentive to hiring young people.

'Bigger changes in attitudes and institutional arrangements are recommended for programs to locate significant portions of education in the workplace. Young people would become part of an organization primarily devoted to work, but in which persons of all ages would have both working and learning roles. The difficulty of incorporating young people would vary, but the panel suggests that it would be possible for them to work in organizations in the performing arts, hospitals, manufacturing and retail businesses, and many government offices. [112]

The panel also recommended consideration of a system of educational vouchers for those over 16, with the vouchers to have the value of the average cost of a college education. As an aside, it is essential to point out that the panel recommends that pilot programs be initiated to test their recommendations and that final policy decisions not be made until the results of the pilot programs are available and analyzed.

Many American citizens are deeply concerned that any kind of major effort to resist inflation will result in increased unemployment among the least advantaged in our society and that the burden of the fight against inflation will be borne heavily by those least able, including those on welfare.

The problem is complicated because our present welfare system has a variety of shortcomings:

—*Welfare recipients frequently receive more income from their welfare benefits than non-welfare families who are working full time.*

—*Benefit levels vary greatly from state to state.*
—*In 26 states, male-headed families generally are ineligible for benefits, even if their total family income remains far below the welfare program's income eligibility criteria.*
—*The rates by which welfare benefits are reduced as earned income increases are frequently so high that a family is discouraged from attempts to supplement welfare benefits by working.* [113]

In 1969 President Nixon sent to the Congress a welfare reform plan, generally associated with the sponsorship of Dr. D. P. Moynihan, [114] which was designed to provide income assistance to all poor families with children, move toward the equalization of benefits among states, assure that work effort would be encouraged, not discouraged, and provide assistance to the working poor.

The plan was approved in the House of Representatives and nearly approved in the Senate, but was killed there finally for widely differing reasons—some political but many also based on concern that any such assistance program would encourage families to rely on the income assistance and withdraw from the labor force. These feared a substantial reduction in the family incentive to work and increasing escalation in the cost of providing benefits.

If a program of this general kind could be installed in the United States without the negative effects, it would go a long way toward alleviating concerns about anti-inflation efforts. Results from an Office of Economic Opportunity experiment launched in 1968 are of particular interest to researchers and policy-makers as they consider welfare reform. The experiment is testing the impact of an assistance system, in many ways similar to the President's program, on a broad variety of issues: work incentive, cost of benefits, administrative costs, and a number of corollary issues such as the impact on health, borrowing and spending behavior, family stability, general attitudes toward work, children's school performance and social behavior, and leisure-time activities. The central objective of the experiment, however, is to determine the relationship of labor supply to the level of benefits and the tax rate on earned income. [115]

one hundred twenty-nine

The experiment is structured to provide assistance that increases as earned income declines and decreases as earned income increases. It differs from the program sponsored by President Nixon in that it does not include a work requirement. It is being conducted in an urban environment and involves a random sample of poor and near-poor families in New Jersey, all with these three characteristics:

—*At least one man (usually the family head) between the ages of 18 and 58 who is neither disabled nor in school.*
—*At least one other person in addition to the family head, i.e., a child, a wife, or an aged relative.*
—*Income in the year before the experiment started not in excess of 150% of the poverty line. (At the start of the experiment, this poverty line was $3,300 for a family of four.) This group is highly significant for policymakers, since the urban, working poor represent about 45% of the families who would be eligible for the Welfare Reform Program. Furthermore, it is among this group that any work disincentive precipitated by an income assistance program would be most likely to be observed.*[116]

A total of 1213 families are involved: 724 receiving the experimental payments and 489 in the control group. Obviously, the experiment is far from complete and findings hardly can be considered as conclusive. However, of the three analyses conducted to date, all draw substantially the same conclusion: "There is no evidence indicating a significant decline in weekly family earnings as a result of the income assistance program." The evidence available thus far indicates that family earnings of the experimental group have not fallen relative to those of the control group.

There also is at least an implication in the data available that this kind of income assistance system may give poor people, especially the working poor, the ability to seek out better jobs. Their dependence on the vicissitudes of low-wage labor markets are reduced, because when faced with unemployment, they are better able to search for higher paying, more permanent employment. If so, this would indeed be an important step forward in policies for dealing with poverty.[117]

In the fight against inflation, which is the common and the

major economic problem of the mixed economies, one of the best weapons is the maximization of the division of labor on a worldwide basis. For nearly a quarter of a century after the ending of World War II, the entire free world, led by the United States, gradually relaxed the trade barriers. An increasingly free trade was accepted as an overall desirable end. In the last few years, nationalism has begun to reintroduce itself, and there are more threats that government interventions around the world will restrict trade to the point that it may upset the dynamic balance of the world economy and push us all into a recession. Here in the United States both organized labor and business groups that find themselves unable to compete effectively with imported products (often because the wage rates paid in the exporting countries are appreciably lower than in the United States) have sought to impose barriers of all kinds.

The Burke-Hartke Bill, which has been discussed extensively in Congress and has had the strong support of organized labor and of a few business groups, is an example. Enactment of any considerable part of the provisions it includes would inevitably decrease the overall competitiveness of U.S. industry and increase the inflationary pressures. It is tragic, indeed almost inconceivable, that at the very time when inflationary pressures are such a severe problem for the mixed economies, there should be serious consideration of any moves except those which would improve trade and maximize the overall use of these countries' resources.

Figure 45 breaks down the structure of the world economy in a somewhat different way from that in Figure 21. Note that the United States, Canada, Japan, and Western Europe (the tripartite countries) with 16.7% of the population, actually generate 60.3% of the world's GNP and just under 70% of total world trade.

This group of countries includes the principal mixed economies of the world. They trade heavily with one another, have similar economic systems, problems, and strengths, along with a considerable commonality of political philosophy. I would suggest that the United States interest a number of these countries in sponsoring a convention to include all of the tripartite nations interested in joining, with the purpose of establishing a general agreement similar in intent and spirit to the General

STRUCTURE OF THE WORLD ECONOMY
1970

	GNP (BILLIONS OF U.S. $ EQUIV.)	% OF WORLD GNP	% OF WORLD POPULATION	% OF WORLD EXPORTS	% OF WORLD IMPORTS
United States	977	27.5	5.6	13.7	12.9
Canada	87	2.4	0.6	5.4	4.4
Western Europe	838	23.6	7.7	44.4	46.3
Japan	245	6.9	2.8	6.2	5.7
Total for Tripartite Countries	2,147	60.3	16.7	69.7	69.3
Other	1,413	39.7	83.3	30.3	30.7
TOTAL	3,560	100.0	100.0	100.0	100.0

Figure 45[118]

Agreement on Trade and Tariffs (GATT), but expressing as a philosophical base a commonality of purpose with respect to commitment to full employment without inflation and improvement in the standard of living with full implication of quality of life, not only for their own citizens but for those of the entire world. This General Agreement for Economic Development (GAED) would not only serve as an overall agreement into which such existing conventions as GATT and those on monetary affairs would couple, but also would go further.

For example, I would suggest:

1. A convention on corporate law, which could simply specify the few major requirements that must be present in the corporate law of the state under which a multinational corporation is chartered. Compliance with this convention on international corporate law then will automatically allow a multinational, based in any of the subscribing countries, to organize a subsidiary in any of the other countries subscribing.

2. The principle of maximizing the division of labor and hence the utilization of resources, both among the subscribing countries and their trading partners in the rest of the world, would be emphasized. Recognizing that adjusting the

economic structures of the various countries will take time, if shocks to certain industrial and agricultural sectors in the individual countries are not to be too severe, general rules covering the establishment of quotas and their phasing out over some maximum amount of time will be included. Under this provision, for example, quotas may be established, when sought by the host country, for industries threatened by imports. Such an adjustment process might have been appropriate for the U.S. textile industry a few years ago when "voluntary" quotas with Japan were arranged. Under this same provision, those countries in Europe with protective walls around their agriculture would agree to similar quotas to be phased out over a span of time.

3. Recognizing the special needs of the developing countries and in fulfillment of the principle of maximizing the division of labor, special encouragement will be given to imports from the developing countries and to the utilization of one of their greatest resources—manpower. Accordingly, for any Less Developed Country (LDC) wanting to participate in the improved tariffs or quotas offered, the subscribing countries may encourage their multinational corporations to establish subsidiaries in these LDCs seeking to maximize the resources available in the LDC, including their labor supply.

4. The General Agreement for Economic Development will include provisions recognizing the right of individual nations to protect, within limits, industries required for the security of the nation. It also will provide for quotas protecting new industries required by a nation to ensure its development and economic stability, but it will adhere to the same principle of phasing out of the quotas over a maximum amount of time.

5. In recognition of the commitment of each nation to full employment, whenever the unemployment rate (calculated according to an agreed-upon method) of any of the subscribing countries exceeds an agreed-upon level, the further phasing out of quotas restricting imports into the country may, on the request of the country, be suspended until the unemployment rate has again fallen to the agreed-upon maximum.

6. The subscribing nations will agree to an enlarged and coordinated effort to aid the LDCs with technical assistance, loans and gifts as appropriate and desired, and always in a

manner designed to advance the policies expressed in this agreement for achieving the productive society, including improved quality of life.

This General Agreement on Economic Development type should be acceptable both to organized labor and to those industries in this country which, for one reason or another, feel themselves non-competitive. It would assure full recognition of the continued emphasis on full employment and, hence, the continued use of fiscal and monetary policy to attain it and relate the quota system both to a phasing-out time and specified unemployment rates. It generally would satisfy those who believe in free trade because of the emphasis on division of labor and maximum utilization of world resources and the provision for a systematic phasing out of restrictions.

Finally, it would be of positive benefit to the citizens of all of the countries because it would maximize resources, provide them with products and services at lower costs, and actively seek for them an enhanced quality of life.

Here in the United States, opposition from organized labor to entering into this kind of General Agreement for Economic Development also should be reduced because labor rates overseas have been inflating at a much higher rate than they have here, and the exchange rate adjustments over the past several years have readjusted the imbalance further.

For example, Figure 46 illustrates in the countries' local currencies the rate of increase in comparable hourly labor costs, including fringes of the country, for our Texas Instruments plants in these countries. Note that the United States, even though it has had increases of approximately 35% between 1968 and 1973, is the lowest of the countries listed, with Italy and Japan showing the highest rates of increase and each more than doubling in that span of time.

Figure 47 compares these same labor rates, including the fringes applicable in the country, converted to dollars, and using our Dallas labor rates in 1968 as 100. It may be surprising to learn that these rates in Germany and France now are ahead of those in the United States, that those in Japan are next highest at nearly 85% of those of the United States, and that the lowest rates among these developed countries prevail in England, about half those of the United States.

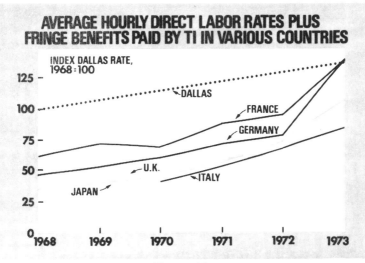

Figure 46

Thus, for a General Agreement for Economic Development, the time is an especially propitious one for the United States because high labor rate differentials, which might have made such an agreement politically impossible earlier, no longer exist; and propitious for the other countries because they are

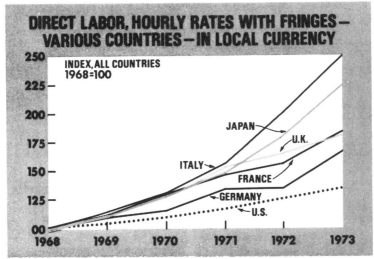

Figure 47

one hundred thirty-five

suffering far worse than is the United States from the pressures of inflation and have even more to gain from their alleviation.

From the standpoint of the developing countries, the gains made in improved trade could be significant. Furthermore, the encouragement of the installation of manufacturing plants, capitalizing on labor-intensive industries in these countries, is remarkably free of most of the emotional overtones that come from the development of natural sources, the extractive industries or the non-labor-intensive industries. Manufacturing know-how and industrial technology are transmitted into the rest of the local industrial effort and local subcontractors inevitably develop. Wage rates will increase with demand, and as the real incomes in these countries increase, so will their abilities to consume. In turn, additional local industries will develop to satisfy these increasing consumer needs and, as the economies of these nations develop, they will provide ever-increasing markets to tripartite countries.

All these countries within the tripartite agreement are high-labor-cost countries and would face generally the same problems of competition and adjustment.

For example, to the extent that shifts of labor-intensive operations take place, the companies within the developed tripartite nations will need to emphasize and increase their efforts to enlarge the competencies of their workers and the jobs they hold. A combined work/education effort will be required, coupled into the needs of all the employing organizations so the education and training given more nearly match the developing needs. The United States, if the combined work/education system described earlier were in place, could use educational vouchers to accomplish retraining of workers in all industries where quotas were in effect, simultaneously contributing to the retraining of the workers and their personal enrichment in a tightly coupled way.

An adequate work/education program will help to reduce inflationary pressures by assuring a supply of trained people for promotion to higher levels. This is to the advantage of the individual as well, since his interest is in a higher real rate of pay, and it is to his benefit to gain that higher real rate, not only through improving productivity on his present job, but wherever possible, by moving on up into a higher level job.

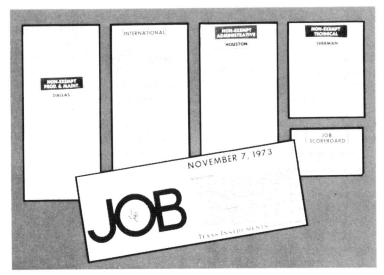

Figure 48

Typical of the kind of mechanism used by many companies but which could be used by more is what we call the TI Job Opportunity System. (See Figure 48.)

TI's Job Opportunity System posts jobs available everywhere in the world in a help wanted-type format inserted in various TI publications. In a recent issue, for example, I noted opportunities for clerks in Houston; technicians and draftsmen in Sherman, Texas; engineering aides, grinder operators, and photomask inspectors in Dallas; a manufacturing supervisor in El Salvador; a manufacturing superintendent in Brazil; and a general foreman in Malaysia.

In the United States alone this year, more than 3000 men and women out of a total U. S. population of approximately 40,000 will have been promoted or transferred to a job they preferred under this system.

An organized work/education program, coupled into this kind of job opportunity system, certainly also would help dissipate the job "blahs" to whatever extent they really exist, discussed with concern in the Health, Education, and Welfare task force study, "Work in America."[119]

At TI we have a program to force us to think hard about all TI employees, right down to the employee on the assembly line, and about their ability and willingness to produce. We call it our

one hundred thirty-seven

People and Asset Effectiveness Program. Through it, we coordinate all of our efforts to improve both productivity and the individual's personal skills and job satisfaction. Included are a wide variety of capital enhancement and automation efforts, many of them heavily dependent upon our increasing ability to utilize computers. At least as important are team improvement groups operating throughout the world to improve productivity on their own jobs. These teams consist of natural working groups that measure themselves on different parameters associated with their jobs in such categories as productivity, material usage, and quality.

For example, in one of our European plants with several thousand employees, there are about 800 women performing semiconductor assembly operations. Of these 800, more than 500 have been involved in these working group programs since the effort was begun in this plant in 1972. Overall, there has been about a 40% improvement in productivity since the beginning of the program, including a decrease from 6% to 7% defective at inspection to below 2%.

In a similar group in Japan, involving what we call bonding semiconductor devices, overall productivity has improved 92% since installation of that program in January 1972.

These teams or working groups exist everywhere, not only in the factory direct labor areas. For example, in the same TI plant in Japan, one work team in our finished goods stockroom improved its productivity 100% in the past year.

In Dallas, a religious holidays committee made up of both salaried and hourly-paid TIers met to recommend pay procedures for religious holidays of minority groups, and a policy was established following the committee's recommendations.

Not surprisingly, every indication we have is that workers participating regularly in such team efforts are much happier in their jobs, and they enjoy improving their own productivity.

A large part of our business remains suprisingly labor-intensive in spite of extensive capital investments following mechanization programs. As a consequence, one of our principal measures in determining our progress under the People and Assets Effectiveness Program is in terms of sales billed per employee across the world. These have risen from $14,600 in 1970 to $21,100 in 1973. Our goal for the late Seventies, and we do expect to reach it, is $30,000 per person.

A very effective mechanism for aligning individual interest and institutional interest is profit sharing. The Profit Sharing Council of America reports that there are over 100,000 tax-qualified plans in the United States. The Council estimates there are an equal number of non-qualified (that is, cash) plans.[120] Many of these plans make their awards in stock of the corporation itself, or the plan buys stock on the open market and/or allows the individual to take his award in stock instead of as a general holding in the total assets of the profit-sharing trust. There hardly can be a better way of aligning the individual employee's and the corporation's interests than a successful profit-sharing plan assuring the individual a considerable stake in the common stock of his own company.

PERCENTAGE OF PROFIT SHARING TRUST ASSETS REPRESENTED BY OWN COMPANY STOCK

	PERCENT	TRUST ASSETS (MILLIONS OF $)		PERCENT	TRUST ASSETS (MILLIONS OF $)
Sears, Roebuck	85.9	3,843	Motorola	14.7	159
J. C. Penney	62.0	409	Bank of America	99.1	142
Standard of Cal.	98.7	281	Chase Manhattan Bank	4.9	140
Eastman Kodak	91.0	257	Zenith Radio Corp.	34.3	139
Federated Dept. Stores	73.6	218	Time, Inc.	N/A	123
Xerox Corp.	6.8	208	Kellog Company	2.1	116
Halliburton Ind.	47.3	179	McGraw-Edison	76.6	115
Burlington Ind.	66.0	181	Texas Instruments	40.8	73
Jewel Companies	7.6	161	Mfr.'s Hanover Trust	80.6	58
Lowe's Companies	87.4	161	Signode Corp.	30.7	53
				AVG. 71.3	TTL. 7,016

Figure 49[121]

Figure 49 shows 20 plans with total assets of $7 billion; 71% of those assets, or $5 billion, was in the form of the profit-sharing companies' own stock. Both the total value and the high percentage are heavily influenced by the profit-sharing plan of Sears, Roebuck and Co. with total assets over $3.8 billion, nearly 86% invested in Sears stock. Sears began this plan in 1915, and today Sears employees, via this profit-sharing trust, own approximately 22% of Sears common stock. The typical retiree

one hundred thirty-nine

with an average service of more than 25 years with Sears received approximately $68,000 from profit sharing in 1971. I would like to emphasize that at Sears, as at most profit-sharing companies, this average of $68,000 is in addition to a normal retirement pension.

Our own TI plan was installed in the 1940s but was only moderately successful until we revised the plan and its sharing formula in 1950. Since our sales billed were only a little over $7.5 million in 1950, we have been a very small company over a long span of the intervening 23 years, so we haven't had nearly the time to generate the kind of strength for our participants as has Sears.

Nevertheless, total assets of our Texas Instruments plan were $73 million at year-end 1972, and through it, nearly 7300 TIers own more than 486,000 shares of TI stock worth over $47 million.

Figure 50 is a list of the profit-sharing dollars received by nine rather typical recent retirees with more than 20 years of service. Since we were pretty small 20-plus years ago, we don't have many retirees who have accumulated more than 20 years of service.

Nine retired at the average age of 63, after working a little more than 25 years for TI. Earnings in their last year were between $6600 and $14,000, averaging about $10,000 per year; 64% of their profit-sharing accounts were in TI stock, and the average total value of their accounts at retirement was more than $59,000, or about six times their earnings in their last year of work. As at Sears, this was in addition to retirement income from a normal retirement plan plus Social Security.

In particular, note the third retiree from the top. Mary B. retired at age 62, after 22 years with TI, with average annual earnings over that time of under $6300. Yet, her profit-sharing account at retirement was more than $50,000, all in TI stock.

I include all of this detail on profit sharing in order to emphasize that, along with my urgent recommendation that we actively seek to establish laws and modes of operation to stop the growing percentage of our total workers employed in the non-profit sector, and especially by government, goes an equally strong recommendation that mechanisms such as profit sharing be adopted broadly to ensure alignment of individual and institutional interests. These kinds of mechanisms share the

EXAMPLES OF 1973 PROFIT SHARING PAYMENTS
TO TI RETIREES WITH 20 + YEARS OF SERVICE

	RETIREMENT AGE	YEARS OF SERVICE	ANNUAL SALARY $	PROFIT SHARING VALUE $	% TI STOCK **
Jean J.	63	21	6,618	36,296	28
Jim S.	62	20	9,378	16,494*	41
Mary B.	62	22	6,277	50,112	100
Tom M.	65	20	7,488	36,367	74
Paul X.	62	26	9,170	35,856*	59
Clarence G.	62	21	12,381	49,038*	46
Murray F.	64	41	13,363	122,738	99
Joe B.	65	40	13,418	100,726	60
Bob B.	65	20	14,018	85,415	69
Average	63.3	25.7	10,235	59,227	64

*Relatively large withdrawals made during career.
**Percentage represents % TI stock in profit sharing account at retirement.

Figure 50

organization's success and the productivity gains made possible through the profit system and the private enterprise sector with the individual to a far greater extent than could conceivably be possible in the government sector. Yet, because of the higher overall productivity of the private enterprise sector, society gets its products and services at lower costs plus, via income taxes, about half of the profits earned!

Probably the best way of characterizing our entire orientation at Texas Instruments is to say that it is objective- or goal-oriented. We do our best to install institutional procedures and practices that involve the largest possible number of employees in establishing goals and cooperating in their attainment. We attempt to align individual and corporate goals to the maximum extent possible and certainly do a better job of this than we used to do, just because we are conscious of the need. We have not succeeded in all of these efforts as well as we should because one of the by-products is recognition, sometimes to one's cha-grin, of the degree to which we are failing to attain specified goals or in our aim to align individual and corporate interests to the maximum extent feasible.

Yet, in the overall, the program has been remarkably success-ful. It does follow the same kind of general principles with

one hundred forty-one

respect to enhancing productivity that I believe apply to our total society, and our growing comprehension and application of the principles have played a major role in the company's growing from just above $2 million per year in billings in 1946 to nearly $1.3 billion in 1973, practically all from internal development.

I believe that all of our institutions, both profit and non-profit, require this orientation to goals and increasing productivity. Our economic system does provide opportunities for accomplishment which can't be matched in a centrally controlled economy, but it also enlarges the responsibilities of the myriad of independent institutions through which it carries out its business. They must set objectives that are both meaningful and ambitious, and which do maximize the alignment of the interests of the individual employees, the institutions themselves, and the societies of which they are a part.

We believe at Texas Instruments that the responsibility for this kind of goal orientation begins with our board of directors, and the board itself emphasizes procedures that directors and other key executives must follow to enhance their own effectiveness.

Briefly, these policies and practices require that inside directors be relieved of all of their other duties to devote a principal amount of time completely to their duties as directors. Beginning in 1973, after age 55, every such inside director must retire from his executive responsibilities and take on the same relationship to the corporation as do outside directors. We concluded that no other course could make the inside director sufficiently dispassionate in his judgments and sufficiently independent of the corporation's other executives. His retirement allowance will be 85% of what it would have been, had he continued working to age 65. All inside directors are expected to take on at least one really meaningful outside responsibility in order to contribute to the dispassionateness of judgment we would like them to couple to their detailed knowledge of our own company activities. Outside directors are required to spend a minimum of 25-30 days annually at TI business and most spend more. Compensation is on an annualized basis but averages out at approximately $1000 per day. The entire board is intimately involved in all long-range planning and policy programs, including the responsibilities for goal-setting and measurement of progress toward objectives.

This emphasis on effectiveness continues right through to the principal executives. Beginning this year, our chairman and president must each retire at age 62. Executive vice presidents will be expected to retire shortly after their 55th birthday and certainly will be prime candidates for inside directors.

To indicate the impact of these policies, if our present chairman, president, and three executive vice presidents all had served to age 65, there would have been only three vacancies in these posts in the next 15 years. Under these new policies, there will be at least seven such major promotions over these same 15 years.

I have gone into this much detail on our TI policies to emphasize that we must take steps to assure growth in the private profit-making sector and to avoid the continued increase in the proportion of our total workers employed directly by the government. I believe equally strongly that we must assure effective operation of our private institutions and that it is the boards of directors of these institutions that have the prime responsibility for seeking that assurance. In my opinion, the several judicial decisions over these past few years that have emphasized the responsibilities of directors were appropriate government interventions to improve the functioning of corporations in the private enterprise sector.

To examine another area where government interventions are appropriate, let's return to Denison's conclusions, as shown in the data displayed in Figure 36. This emphasized the increased growth in national income and in national income per person employed resulting from improved knowledge, first, in the form of a more effective labor force through education, and second, through applications of knowledge as expressed in technological innovation and improved management. Earlier, I stressed the importance of education as a source of growth. However, as these data from Denison show, the contribution of knowledge itself as embodied in technological change and improved management plays an even more significant role—about 0.76% contributed by the advances in knowledge against 0.49% from education. Advances in knowledge alone, exclusive of the gains from education, provided 23% of our increased growth in

one hundred forty-three

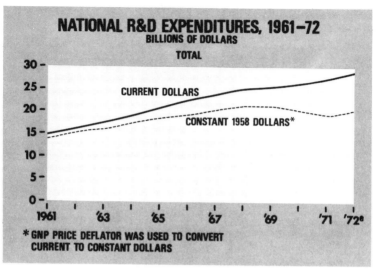

NATIONAL R&D EXPENDITURES, 1961–72
BILLIONS OF DOLLARS

TOTAL

CURRENT DOLLARS

CONSTANT 1958 DOLLARS*

30 –
25 –
20 –
15 –
10 –
5 –
0 –

1961   '63   '65   '67   '69   '71 '72ᵉ

*GNP PRICE DEFLATOR WAS USED TO CONVERT
CURRENT TO CONSTANT DOLLARS

Figure 51[122]

national income and 35% of the increase in national income per person employed.

That being the case, one cannot help wanting to examine carefully some recent trends in overall research and development efforts in the United States. Research and development contributes only part of the advances in knowledge to which Denison assigns such an important place as a source of growth. Nevertheless, R&D is a principal component and one for which we can measure the aggregate input (Figure 51).

Our national R&D expenditures, expressed in current dollars, grew steadily to a peak of about $29 billion in 1972. Expressed in constant 1958 dollars, however, they reached their peak in 1968 at about $20.5 billion and then declined about 6% between 1970 and 1971, increasing again slightly in 1972, to about $20 billion 1958 dollars, or about the same as we were running back in 1966 and 1967.

Furthermore, the number of scientists and engineers employed in research and development climbed to a total of 560,000 in 1969 and fell steadily after that to about 525,000 at the end of 1972, a decline of some 34,000.(See Figure 52.)

Most of this decline in research and development came about because of the cutbacks in federal expenditures for national defense and space, so from the standpoint of contributions to economic growth, it is entirely possible that to the extent this results in a shift in emphasis to industrial and such other areas as health and education, it may in the long run be constructive from a growth standpoint.

When one looks at the source of funds for industrial R&D (Figure 53), there is some evidence that the resources freed by the government cutbacks in defense and space expenditures are to a considerable extent being transferred by industry to expenditures for its own purposes. Thus, while federal support for industrial R&D fell from a peak of $7.3 billion in constant 1958 dollars in 1966 to $5.6 billion in 1972 (or a cut of nearly one-quarter), about $1.5 billion of the $1.7 billion lost has been replaced by funds from industry which should be more effectively focused.

Earlier, I concluded from Denison's study (comparing the sources of growth in seven European countries with those of the United States) that the contributions "from knowledge" were a more important source of growth for an advanced society like that of the United States than for others in an earlier state of development, which could draw upon the knowledge available elsewhere to a considerable extent.

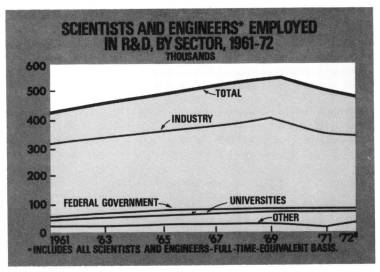

Figure 52[123]

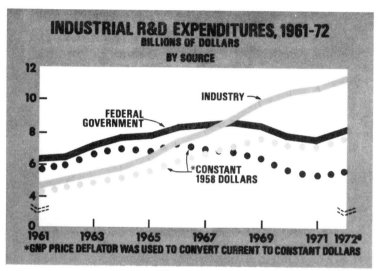

Figure 53[124]

In that sense, I find it is troublesome that the percentage of our gross national product we are spending on research and development has been declining sharply while the percentages of GNP that Japan, Germany, and the USSR are spending have continued to increase. (See Figure 54.) The actual expenditures in both West Germany and Japan (at 2% or below) still are well under ours at 2.6% in 1971. But because these two countries do not have any significant expenditures on research and development for defense or space, their research and development expenditures for all other areas exceed ours as a percentage of GNP.

The expenditures of the USSR, with large defense and space components, actually have crossed ours and are now some 3% of their GNP.

Of even more concern are the statistics on the number of scientists and engineers engaged in research and development per 10,000 population. After a climb to 28 per 10,000 in the United States, our level now has fallen to 25 per 10,000, the same as Japan's and only two-thirds that of the USSR at 37. Further, as Figure 55 portrays, we are the only one of these countries with a declining percentage.

Of course, not one of these gross numbers says anything about the quality of the individual researchers or their work, or

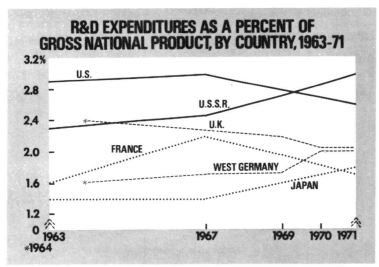

Figure 54[125]

about the effectiveness of the coupling of the research and development into the applications that are necessary if the acquired knowledge is to become a source of growth. It would be my personal judgment, based on our TI experience, that a combination of higher quality work and better coupling to

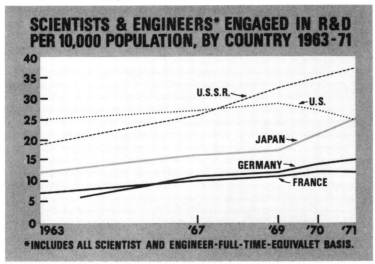

Figure 55[126]

one hundred forty-seven

applications can and often does make such R&D expenditures 10 or more times as effective as the average.

In commenting earlier on Texas Instruments growth, I emphasized that it had been generated almost entirely internally. A principal component contributing to that internal growth has been innovation in all aspects of the business and emphasizing, not just research and development, but its coupling into the manufacturing and marketing functions.

We have a formal, institutionalized system that provides for deliberate, planned innovation in all our product and service areas. With this institutionalized, long-range planning system we attempt to manage innovation and our whole company to provide continuing stimulus to the company's growth and usefulness to society and as a business institution.

Essentially, the system includes detailed and scheduled short- and long-range objectives for every segment of Texas Instruments, as well as for the company as a whole, for a decade into the future. The strategies we will follow in order to attain the objectives are identified and stated succinctly in writing. The specific tactics that will be implemented in order to pursue the strategies are detailed. Tactics may be specific research and development programs to provide a new product, to solve a process problem, or to identify an opportunity in an exciting new technical field. They include automation programs, personnel programs, and market implementation programs as well. We call this overall, formal system our Objectives, Strategies, and Tactics (OST) System.

Each year, in implementing it, we target sales billed and manufacturing and operating cost goals and determine how much will be left over to cover our total expense package plus profits. Expenses are in two categories: first, our operating expenses which relate simply to maintaining our present business; and second, discretionary expenses related to growth and development of the company. If one is willing to forego growth in quality and size, discretionary expenses, unlike operational expenses, can be omitted or delayed. We make a deliberate choice as to how much to invest in discretionary expenses to support these OST programs, taking into consideration the state of our business and the profit level we feel we must sustain now in order to be able to exercise the opportunities generated in the future.

To assist in making these choices, each package of activity on which an independent decision logically could be made—which we call a "decision package"—is analyzed, scored on a set of criteria, and ranked. (Figure 56.) Each program manager scores his own program. It then is reviewed and ranked at successively higher levels—strategy, objective, and corporate management. The decision packages then are sorted and evaluated by strategy, by technology, and by area of business. We are attempting to minimize duplication of effort across the organization, to make sure that we are not overemphasizing slow-growing aspects of the business to the detriment of faster-growing ones, and to select the best mix between low-risk, low-payoff and high-risk, high-payoff activities.

From the rankings we can determine which decision packages can be undertaken through any particular level of discretionary funding. We are always able to define solidly more decision packages than we can implement. This limitation is imposed both by the funds and qualified people available. The difference

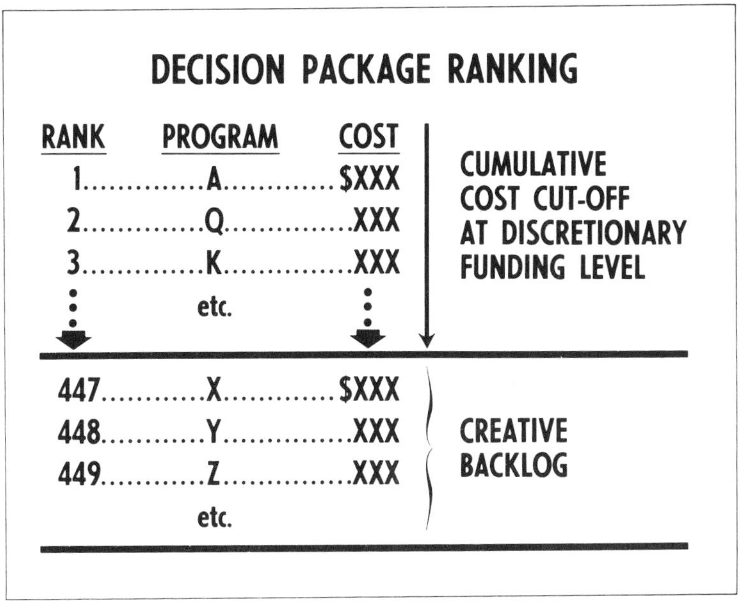

DECISION PACKAGE RANKING

RANK	PROGRAM	COST	
1	A	$XXX	CUMULATIVE COST CUT-OFF AT DISCRETIONARY FUNDING LEVEL
2	Q	XXX	
3	K	XXX	
⋮	etc.	⋮	
447	X	$XXX	CREATIVE BACKLOG
448	Y	XXX	
449	Z	XXX	
	etc.		

Figure 56

between the identified programs and the actual programs undertaken is called a "creative backlog." Many of the programs in the backlog are undertaken later as resources become available. You will recognize that this is a species of zero base budgetir.g, a technique that was originated at Texas Instruments several years ago, in the development of which Peter A. Pyhrr played a major role and described in his book, *Zero Base Budgeting*.[127]

Many companies in private industry have mechanisms to assure that their innovative efforts, and especially their research and development, are properly coordinated and applied as parts of an overall plan. Unfortunately, even more do not. And rarely in my rather broad experience with non-profit organizations, both in government and outside of government, including universities, have I ever found even a rudimentary equivalent of this system that requires that we state our objectives, strategies, and tactics for development and growth, structure our management and accounting systems to pursue them, and measure progress or failure regularly.

Thus, although I am concerned about our overall decline in research and development in the United States, I am even more concerned about the absence of effective mechanisms for utilizing the research and development, and my concern is amplified because of the shift in our society toward more and more activity in the government and other non-profit sectors.

Edward Mansfield, in his paper "The Contribution of Research and Development to Economic Growth in the United States,"[128] comments that two propositions bearing on U. S. investment in research and development are widely accepted by economists. These are that there is good reason to believe that, if left to its own devices, the market would allocate too few resources to R&D—and that the short fall would be particularly great at the more basic end of the R&D spectrum, because (1) the results of research are often of little direct value to the sponsoring firm but of great value to other firms; and (2) research and development is risky for the individual firm.

The reasons are obvious. The market operates on the principle that the benefits go to the person bearing the costs and vice versa. If a firm or individual takes an action that contributes to society's welfare, but it cannot realize sufficient gain, it is less likely to take this action than would be socially desirable. Further, the risk to the individual investing in R&D is greater than the risk to society since the results of R&D may be useful to someone else, not to himself, and he may be unable to obtain from the user the full value of the information. Thus, because the economic system has limited and imperfect ways to shift risks, there would be an underinvestment in R&D.

Where the private business firm takes advantage of its much greater ability to achieve proper coupling between research and development and its other activities, I believe that this tendency to underinvest does not exist except as one approaches the basic research end of the spectrum. Unfortunately, business organizations that use research and development effectively, while more common in the United States than elsewhere in the world, still are not as numerous as they should be, and they generally are found in the progressive and high-growth sectors where this ability probably is both cause and effect.

As a consequence, it is probable that the Federal Government should get more deliberately and deeply involved in private sector research and development under certain conditions. For example:

1. When the results of R&D are likely to have major beneficial effects on the welfare of a large segment of the population (e.g., agriculture in the past, perhaps housing in the future, energy right now) and when one of the following additional conditions holds true:

   a. The private sector is too fragmented for any single entity to be able to finance a meaningful feasibility demonstration;

   b. The risks of failure, the length of the effort, and the total size of the investment, combined with the relatively small fraction of the resulting economic value that might accrue to the developer, contribute to the discouragement of private investment (e.g., nuclear reactors, high-speed transportation);

   c. Social returns appear to be so high that acceleration is required. The present energy crisis illustrates such an opportunity for R&D in energy sources as alternates to petroleum;

2. When there is a set of phenomena newly discovered or not yet sufficiently explored (e.g., high-energy lasers) where

one hundred fifty-one

basic research investigation appears likely to lead to new insights of significant importance to the government or to the possibility of creating a wholly new industry;

3. When a great variety of payoffs is expected, but R&D is not profitable to individual firms because of the narrowness of their product interest relative to the range of R&D application (e.g., social research on cities);

4. When the generation of the requirements themselves requires substantial investment in R&D. This is particularly true of completely new technologies the applications of which are not easily foreseen (e.g., nuclear energy, space, high-power lasers).

Wherever there is federal involvement in private-sector R&D, adequate mechanisms should be developed for the full and extensive exploitation of the fruits of R&D. For example, built into the mechanisms for federal support of R&D should be the means for expeditious and timely transfer of the effort as well as the funding responsibilities, when it becomes appropriate, to mission-oriented agencies, state governments, and private companies.

Because of the extreme importance of adequate coupling, however, any public policy with respect to increasing federal expenditures in industrially directed research and development must not be constructed simply to get industry to do more R&D. Instead, that public policy must seek to improve the industrial objectives themselves and increase the probability of their being attained more effectively and more profitably through well-conducted and well-coupled research and development.

On the other hand, I am convinced of the complete appropriateness of Mansfield's two propositions in the basic research area. An overwhelming proportion of the support for basic research simply must come from Federal Government sources. Consequently, the fact that the same kind of decline exists in real expenditures for basic research is cause for concern.

Our basic research spending in 1972 in constant 1958 dollars amounted to $2,812,000,000, again down about 6% from the 1968 peak of $2,991,000,000 and not too different from our 1966 level (Figure 57). The Federal Government supplied 62% of total funds for U. S. basic research and it is the cutback in federal funds that has been responsible for the overall decrease. In-

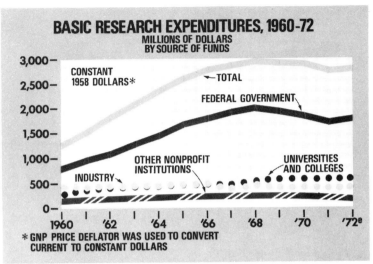

BASIC RESEARCH EXPENDITURES, 1960-72
MILLIONS OF DOLLARS
BY SOURCE OF FUNDS

CONSTANT 1958 DOLLARS*

TOTAL

FEDERAL GOVERNMENT

OTHER NONPROFIT INSTITUTIONS

UNIVERSITIES AND COLLEGES

INDUSTRY

*GNP PRICE DEFLATOR WAS USED TO CONVERT CURRENT TO CONSTANT DOLLARS

Figure 57[129]

deed, the universities and colleges, through their own funds, have made up to a very limited extent for the federal decrease.

Because I am convinced that basic research is a vital element in both our cultural and economic development and because the majority of the support must come from the Federal Government, I am disturbed by these reductions. At the same time, I do not believe it is at all clear that the overall accomplishments of academic science correlate well with the enormous increase in support since 1955. For example, science advocates inevitably emphasize their need for funds being increased by an amount each year sufficient to overcome inflation and add real growth. Practically, and only very rarely, is there any kind of admission that a deliberate effort at improving productivity might overcome at least the inflation. This is particularly serious in academic science, because the great pressures created by the extraordinary rate of growth since the Middle Fifties also have produced a rate of inflation well above that of the general economy. So, without significant improvements in productivity, the real funds available for academic science undoubtedly are well below those presented in terms of constant 1958 dollars.

It is not easy, of course, to establish productivity in academic science because the specific consequences of its efforts are so

one hundred fifty-three

unpredictable; yet, I continue to believe we need a really comprehensive overall survey that will penetrate and illuminate the extraordinarily diffuse structure of academic science. Such a survey would have to penetrate the hundreds of institutions, many thousands of programs and tens of thousands of individual investigators in sufficient depth and with sufficient coherence that conclusions can be drawn both as to the effectiveness of the program as presently structured and how it has been conducted over the past several decades, but most important, so we can determine how to make it better.

I welcome the report from the National Science Board, "Science Indicators 1972," from which most of the R&D statistics I have discussed with you have been taken. It is, as NSB chairman Carter emphasizes, only a beginning—the first word, not the last—in a long-term effort to develop reliable measures of the health of science. I certainly would hope that as a part of that effort, the kind of in-depth survey I have recommended could be included.

Also, in a report issued in 1971 by the National Science Foundation[130] in response to a request by the Office of Management and Budget, attempting to relate research and development and economic growth, there are innumerable recommendations on a variety of economic research programs badly needed on the research process itself and its coupling to economic growth. I hope that the National Science Foundation, as a part of this long-term program to measure the health of science, will begin, too, to include the kind of research suggested by Mansfield and others in this 1971 report.

Until early 1973, there was in the Executive Office of the President a science advisory mechanism that included the President's Science Adviser, the President's Science Advisory Committee (PSAC), and the Office of Science and Technology. In January 1973, the Office of Science and Technology and PSAC were eliminated and the remaining functions transferred to the National Science Foundation. Dr. Guy Stever, the Director of the National Science Foundation, also was made Science Adviser.

Certainly, many of the science-and-technology-related activities of significance to the Executive Office of the President can be handled as well or better with this new mechanism, but I am convinced that science and technology contribute so vitally to economic welfare, both as key sources of economic growth and in determining appropriate government intervention, that the complete elimination of the science advisory mechanism from the Executive Office of the President would appear to have been unwise.

I suggest that there is one set of actions we could take as a nation that would provide an overall framework for the variety of decisions we must make to become a more productive society, and at the same time restore to the Executive Office of the President a competence in science and technology in a manner that would allow it to relate significantly to national policy. This set of actions is described in Chapter IV.

The Productive Society

IV. The National Development Act of 1976

IV. The National Development Act of 1976

Following the extensive debate on the economy after World War II, Congress enacted the "Employment Act of 1946." That historic act created the Annual Economic Report of the President, the Council of Economic Advisers, and the Joint Economic Committee of the Congress.

In its declaration of policy:

The Congress declares that it is the continuing policy and responsibility of the Federal Government to use all practicable means consistent with its needs and obligations and other essential considerations of national policy, with the assistance and cooperation of industry, agriculture, labor, and state and local governments, to coordinate and utilize all its plans, functions, and resources for the purpose of creating and maintaining, in a manner calculated to foster and promote free competitive enterprise and the general welfare, conditions under which there will be afforded useful employment opportunities, including self-employment, for those able, willing, and seeking to work, and to promote maximum employment, production, and purchasing power.

I suggest, there would be no more fitting way to celebrate the 200th birthday of this nation than with a new act, the "National Development Act of 1976," as a natural evolution from that "Employment Act of 1946."

As I envision it, this new National Development Act would declare that it is the continuing policy and responsibility of the Federal Government:

1. To seek for every citizen an ever-improving standard of living defined in the full context of quality of life as well as material affluence;

2. To encourage all practicable means to foster and promote free, competitive enterprise to fulfill needs of our citizens for goods and services;

3. To use government intervention, but with caution and understanding, to modify the market economy for goods and services, to affect the price structure of goods and services so they reflect the value of such public goods as the environment, or to impose overall regulation where the welfare of

society (such as for health or safety) is concerned;

4. To require modes of federal intervention that will avoid government ownership of facilities, and minimize direct government employment of workers;

5. To use all practicable means consistent with its needs and obligations to assure that there will be useful employment opportunities, including self-employment, for those able, willing, and seeking to work and to promote maximum employment, production, and purchasing power;

6. To conduct its affairs and interventions so as to provide a stable and growing economy with a minimum inclination to inflation;

7. To encourage a broad enlargement of educational opportunities with emphasis on equal opportunities for all men and women throughout life, including especially combined work-learning programs aimed at consistently upgrading the skills of workers everywhere and to their broad cultural betterment;

8. To foster the growth of knowledge throughout the society in all fields, including science and technology, art, and the humanities, with particular emphasis on those basic areas vital to the continued economic growth and social development of the United States, including the use of research and development as key tools in attaining national objectives.

9. To seek improvement in the standard of living (with full implications of quality of life) of the world's peoples by joining in cooperative agreement and efforts with other nations, especially those whose political and economic philosophies are compatible, which will advance the policies expressed in this Act throughout the world, and:

a. enhance free trade, travel and commerce,

b. assure sound international currencies,

c. minimize environmental pollution,

d. improve the health and protect the safety of all peoples,

e. maximize the utility and conserve the resources of all nations,

f. encourage competitive enterprise (including multinational corporations) to fulfill needs for goods and services.

one hundred fifty-nine

Under the Act, the President will transmit to the Congress in January of each year a National Development Report reviewing the overall quality of life in this country, including not only the overall economic performance in such usual parameters as GNP and NNP but also a broad variety of such necessary aspects of quality of life as an improved environment and educational and cultural attainment. Every effort will be made to encourage and use as they become available such indices as the measures of economic welfare (MEW) proposed by Nordhaus and Tobin so that over the years the report may truly present a comprehensive statement on our national quality of life and programs relating to its improvement.

The Act will establish a Council of National Development Advisers in the Executive Office of the President composed of five members appointed by the President by and with the advice and consent of the Senate, each of whom shall be a person who, as a result of his training, experience and attainments, is exceptionally qualified to analyze and interpret developments in economics, education, and science and technology; to appraise programs and activities of the government in light of the policy declared by this act; to formulate and recommend national development policy to promote employment, production, and purchasing power under free enterprise to resist inflation; and generally to advance the standard of living of all citizens of the United States with full emphasis on quality of life as well as material affluence.

One of the members of the Council will be designated by the President as Chairman of the Council and his Chief National Development Adviser. The complement of five members will be structured to include economists, scientists, engineers, educators, and industrial executives in such a combination as to provide the Council broad knowledge and experience in economic theory and practice, science and technology, lifetime education, and the management of complex enterprises in an international environment as requisite to attainment of the policy objectives of this Act.

The Council will:

1. Assist and advise the President in the preparation of the National Development Report;

2. Gather timely and authoritative information concerning national development trends both current and prospective; analyze and interpret such information in light of the policy declared in this act; compile and submit to the President studies relating to such developments and trends;

3. Appraise the various programs and activities of the Federal Government in the light of the policy declared in this Act;

4. Recommend to the President national development policies to foster the objectives of this Act;

5. Recommend programs to develop measurements which will evaluate more accurately our national standard of living in its full implications for quality of life.*

The Act will establish a Joint Development Committee to be composed of eight members of the Senate to be appointed by the President of the Senate and eight members of the House of Representatives to be appointed by the Speaker of the House (in each house the majority party to be represented by five members and the minority by three).

The Joint Development Committee will:

1. Make a continuing study of matters relating to the National Development Report;

2. Study means of coordinating programs in order to further the policies established by this Act;

3. As a guide to the several committees of the Congress dealing with legislation relating to the National Development Report, prepare annually and file with the Senate and the House a report containing its findings and recommendations with respect to each of the main recommendations made by the President in the National Development Report.

The Act will establish as a matter of policy that the Congress will set annual spending limitations and the maximum budget

---

*See the paper "Measuring, Monitoring and Modeling Quality of Life" by Charnes, Cooper, and Kozmetsky for an excellent discussion of multi-dimensional approaches to measuring "quality of life."[131]

outlays for both the appropriations and legislative committees of the Congress.**

* * *

I hope that the public debate which would accompany the preparation for this National Development Act of 1976 and its enactment would serve to (1) inform the citizens of this country as to the base of their material affluence and quality of life, (2) point to approaches which are most likely to alleviate or remove the very real faults of our society, (3) emphasize and utilize the enormous capabilities of this nation in science and technology through the applications of research and development aimed at the attainment of national objectives, and (4) protect and enlarge the freedom and dignity of every citizen.

Even with the best will and most competent sponsorship, progress toward the overall goals discussed in these papers must be painfully slow, and if the perspective both of our shortcomings and of how well we have really done by comparison with the rest of the world is lost, we are likely also to lose our way toward a truly productive society.

One of our principal dangers to attaining even that slow progress will be the impatience and emotional reaction of those who feel either that no progress at all is being made or that an immediate cure for one of our numerous social or environmental ailments is so vital that any attempt which involves the tedium of analysis, careful judgment, and the balancing of all objectives will be viewed as simply stalling.

The nature of man and his curiosities being what they are, most of the instantaneous vignettes of life gained so readily and speedily through radio and especially television are related to what is wrong in our country and to the troubles of the world.

Furthermore, the very instantaneousness of the transmission of radio and television, the apparent ease with which these eliminations of time and space are accomplished, the apparent

---

**Fundamentally, this requirement is in agreement with H.R.7130 Budget and Impoundment Control Act of 1973 (sponsored by Al Ullman, Representative from Oregon), which as of July 1974 has passed the House and Senate and awaits the President's signature.*

clockwork precision of the landings on the moon, all conspire to give the average citizen an exaggerated impression of the power and the competence of our science and technology and of our nation's executive skills. They tend to convince him that we have both the knowledge and the affluence to solve all our problems if we but had the will to do so. Little wonder then that so many conclude that since we have all of this knowledge and all of this wealth, it must be either the malevolence of men —those who control the establishment—or the impersonal overwhelming clockwork of the functioning machines that keeps us from attacking and solving all these problems that prevent us from living in a Utopia.

Each of us can extrapolate only from his own experience and knowledge. From my own I have come to a view which seems to me to account for much of the frustration and of the impression that it is futile to attempt to change our institutions. I have come to that view from examination of that which I know best; my own company, Texas Instruments.

Now, by all comparative standards, Texas Instruments must be considered among the better industrial organizations. We have grown from $2 million per year total sales billed in 1946 to almost $1.3 billion per year in 1973. We have contributed a number of products and services of significance, including the silicon transistor, integrated circuits, key methods of oil exploration, and a variety of others of sufficient worth so that they have generated the growth we have enjoyed. We are an organization that emphasizes innovation based on science and technology and with a strong bias toward organized institutionalized management techniques.

Yet, when I compare the programs where we have done best—the programs where our research and development have had a peculiarly effective insight, where the manufacturing processes have been developed and selected properly, where we truly comprehended and solved our customer's problems—I am forced to conclude that, in these program areas we have handled best, we are at least 25 to 50 times more effective than we are on average.

Further, when I examine these best programs, it would appear that they, too, could have been done from five to 10 times more effectively if we had only known enough, or understood the facts of nature, or one another, or our customers' needs

one hundred sixty-three

better. Thus, I conclude that in the overall, Texas Instruments cannot be more than 1% effective as measured against what is theoretically possible.

And yet I know that doing as well as we have has been an extraordinarily difficult task. It has required long hours of effort over decades by hundreds and thousands of key men and women who themselves have been more able than the average; knew, admired, and respected one another; and understood their goals and objectives better than is common in most human institutions. Many of our most successful technical programs have taken longer and have cost more than we anticipated. Some of them have been dangerously close to failure—not once but several times—before final accomplishment. It would not be inaccurate to say that, even with our best efforts, a number of effective programs have barely attained satisfactory objectives.

My experience with other industrial organizations, with the universities I know, and with government, convince me that they, too, are 1% or less effective against any kind of similar theoretical standard. In fact, my own experience would suggest that universities and government have far more potential for improvement than today's better industrial organizations.

This situation can produce the optimism about science and technology and the industrial society that was characteristic up to a few decades ago or the black pessimism that generates some of the kinds of criticisms with which I started this discussion. It all depends on the point of view. If the focus is on how much more effective today's industrial societies are at satisfying man's material wants than were the feudal and pre-feudal societies, or are the completely planned societies or those of today's underdeveloped world, then all of the optimism is justified. If on the other hand, the concentration is on what yet remains to be done, on the potentials which exist, and on how far we are from attaining our ideals, either in satisfying material wants or in moving toward the ultimate in man's spiritual development, then, of course, what we have accomplished seems inconsequential. Of even greater import, the rate at which we are moving toward such perfection will seem imperceptible.

Again, let me turn to Texas Instruments to attempt to make the concept clearer. I believe that our company's sales are a measure of the value of our products and services to society and that our profitability is a measure of how effectively we operate to create, make and market those products and services. On the average, Texas Instruments attempts to make about 10% to 12% profit on its sales billed before taxes, or about 5% to 6% after taxes. Thus, a 10% improvement in profits is attained by a 1% decrease in the overall costs we accumulate in generating our total sales billed. That kind of improvement ordinarily is the result of a large number of deliberate choices, the successful execution of a wide variety of programs, and extremely effective institutional operation over a considerable span of time. In other words, it is the result of a great deal of hard work to attain deliberately and carefully chosen objectives. Attainment of that kind of improvement ordinarily would and should result in a considerable degree of satisfaction. However, if Texas Instruments is about 1% effective and if one is focusing, not on improving the organization as it exists but on all of the opportunity around it, on all the things yet to be done, on all of the errors to be corrected; that is, if one is concentrating on the 99% yet to be done, then that 10% improvement in effectiveness or that 1% decrease in overall cost is seen as a 1/100% gain against what yet remains as possible. In fact, if one is concentrating on what could be done, it is highly probable that it will appear that the gains are lost in the noise, no progress has been made, and all of this effort has been futile.

If I am evaluated as a professional and an executive on the basis of what Texas Instruments has accomplished by comparison to its competitors around the world—in other words, in terms of what my associates and I have done right—we must be given relatively good grades. On the other hand, if I am evaluated in terms of what *could* be done, measured on my mistakes—what I didn't see, what I didn't know, what I didn't do—I am a failure.

In the sense that they do recognize the enormous potential for improvement in our society, such criticisms as those I've quoted in this discussion are justified, but rarely have the critics presented their arguments or their solutions in terms that suggest to me that they really understand how difficult it is to improve *any* complex institution.

Reich comes very close to describing the dilemma without,

it seems to me, understanding it in these paragraphs from *The Greening of America*:

There is a great discovery awaiting those who choose a new set of values—a discovery comparable to the revelation that the Wizard of Oz was just a humbug. The discovery is simply this: there is nobody whatever on the other side. Nobody wants inadequate housing and medical care—only the machine. Nobody wants war except the machine. And even businessmen, once liberated, would like to roll in the grass and lie in the sun. There is no need, then, to fight any group of people in America. They are all fellow sufferers. There is no reason to fight the machine. It can be made the servant of man. Consciousness III can make a new society.

'The crucial fact to realize about all the powerful machinery of the Corporate State—its laws, structure, political system—is that it possesses no mind. All that is needed to bring about change is to capture its controls—and they are held by nobody. It is not a case for revolution. It is a case for filling a void, for supplying a mind where none exists. No political revolution is possible in the United States right now, but no such revolution is needed.[132]

Clearly, Reich is looking at the 99% we still can accomplish, but he has not appreciated how difficult it has been to get this far and how much diligent effort it will require to continue to gain from where we are.

In that sense, given the complexities of our society and the problems we face, I know the recommendations I have made are anything but a complete prescription for the country's future. Yet, they do reflect my fundamental optimism as to the wisdom of our citizens and the overall decency of the governments they elect. But it will take a very wise people to select and pursue a proper course toward an improved society. I do believe that the constructive part of the current criticism in our society has contributed to our becoming a wiser people. We are learning that our standard of living is lower when we allow racial prejudice to blight our society or when we lose our forests and our streams or when the air in our cities is so polluted we no longer can enjoy our surroundings. This is a large gain in wisdom. Yet, it will take a very wise people indeed to make the choices that will allow us to attain the truly productive society and improve the overall quality of our life without simultaneously destroying the only system which thus far has made such choices feasible.

References

1. Plato's "Discourses on the Republic,"
   "Great Books of the Western World," *Encyclopedia Britannica*, 1952, Vol. 7., p. 316.
2. Constitution of the United States of America.
3. Keynes, John M., *The General Theory of Employment, Interest, and Money.* New York: Harcourt, Brace & World, Inc., (c) 1965., pp. 383-4.
4. Mumford, Lewis, *The Myth of the Machine: "The Pentagon of Power."* New York: Harcourt Brace Jovanovich, Inc., (c) 1970., p. 337.
5. Reich, Charles A., *The Greening of America.* New York: Random House, (c) 1970., pp. 87-88.
6. Skinner, B. F., *Beyond Freedom and Dignity.* New York: Random House, (c) 1971., p. 1.
7. Gray, Francine du Plessix, *Divine Disobedience: Profiles in Catholic Radicalism.* New York: Alfred A. Knopf, (c) 1970., p. 220.
8. Samuelson, Paul A., *Economics*, 9th Edition. New York: McGraw Hill, Inc., (c) 1973. Used with permission of McGraw-Hill Book Co. Texas Instruments Advanced Economic Planning Department.
9. Kuznets, Dr. Simon, *Modern Economic Growth: Rate, Structure and Spread.* New Haven: Yale University Press, (c) 1966., pp. 482-483.
10. Historical Statistics of the U.S., Series D, 589-602, p. 91.
11. Economic Indicators, Council of Economic Advisors, October, 1973.
12. Samuelson, op. cit.; TI Advanced Economic Planning Department.
13. Demographic Yearbook, United Nations, 1971; Statistical Outline of India, 1967; Agency for International Development, 1971.
14. Ibid.
15. Vital Statistics of the U.S., U. S. Public Health Services.
16. Ibid.; 1950-1970 data—Consumer Income Series P-60, #85, 12/72. 1930 data—estimated by Texas Instruments Advanced Economic Planning, using 1929 data from *America's Capacity to Consume*, Maurice, Leven, et al. Brookings Institution, 1934, p. 52.
17. Demographic Yearbook, United Nations, 1971. Gross National Product —Growth Trends, Agency for International Development, 1973.
18. "Digest of Educational Statistics," U. S. Department of Health, Education and Welfare; U. S. Department of Commerce, Bureau of Economic Analysis.
19. Deane, Phyllis and W. A. Cole, *British Economic Growth, 1688-1959* (Cambridge, 1962), Table 28, p. 127.
20. Yasuba, Yasukichi, *Birth Rates of the White Population in the United States. 1800-1860* (Baltimore, 1961), Table III-1, p. 75.
21. Kuznets, op. cit., p. 46.
22. Fuller, H. I., "Cleaner Air in London," *Petroleum Review*, February 1971.
23. Ibid.
24. Smith, Gene (Photographer), *UNESCO Courier.* July 1971, v. 24.
25. Stockton, Edward "Pittsburgh—from Smoky City to Smog-Less Skies," *UNESCO Courier,* July 1971, v. 24.
26. Council on Environmental Quality, Third Annual Report, August 1972.

27. Ibid.

28. Ibid.

29. Ramos, Mike (Illustrator), *Smithsonian Magazine*, May 1973, v. 4, #2, p. 100.

30. Berkebile, Don H., "Get a Horse? Save Us From That Pollution." *Smithsonian Institution*, May 1973, v. 4, #2, p. 100.

31. Ellis, Mel, "The Good Earth," *Milwaukee Journal*, October 28, 1973. Used with permission of Associated Press.

32. *Encyclopedia Britannica*, 1968, v. 22, p. 692.

33. "Digest of Educational Statistics," U. S. Office of Education, 1970.

34. Boorstin, Daniel, *Decline of Radicalism: Reflections on America Today*. New York: Random House, Inc., (c) 1970.

35. Nordhaus, William and James Tobin, "Is Growth Obsolete?" *Economic Growth* (50th Anniversary Colloquium V). New York: National Bureau of Economic Research (General Series 96), 1972, p. 4.

36. Ibid., Table 2, p. 13.

37. Ibid., Table 1, p. 12.

38. Ibid, Table 1, p. 12.

39. Ibid., p. 13.

40. Forrester, Jay W., *World Dynamics*. Cambridge: Wright-Allen Press, Inc., (c) 1971, pp. 11-13.

41. Ibid., pp. 124-5.

42. Solow, Robert M., "Notes on Doomsday Models." *Proceedings of the National Academy of Sciences*, Vol. 69, #12, pp. 3832-3833., December 1972.

43. Ibid., "The Chicken Little Syndrome."

44. Nordhaus and Tobin, op. cit., p. 15.

45. Ibid., pp. 15-16.

46. Ibid., p. 17.

47. Ibid., Table 3, p. 19. TI Advanced Economic Planning Department (1972 est.).

48. "Population Index," April 1973. Office of Population Research, Princeton University.

49. Nordhaus and Tobin, op. cit., pp. 22-24.

50. Fried, Edward R., "Foreign Economic Policy: The Search for a Strategy," *The Next Phase in Foreign Policy*, edited by Henry Owen. Washington, D.C.: Brookings Institution, (c) 1973, Table 3, p. 168.

51. Samuelson, op. cit., p. 769.

52. Galbraith, John K., *Economic Development in Perspective*. Cambridge: Harvard University Press, (c) 1962, pp. 1, 2.

53. "Money Income of Families & Persons in the U.S.," *Consumer Income*, U.S. Department of Commerce Series P-60, #87, June 1973.

54. Samuelson, op. cit.

55. Ibid., p. 883.

56. Sik, Ota, *Czechoslovakia: The Bureaucratic Economy*. White Plains: International Arts & Sciences Press, Inc., (c) 1972, p. 76.

57. Ibid.

58. Ibid., Diagram 3, p. 65.

59. Ibid., pp. 66-67.

60. Ibid., p. 101.

61. Erhard, Ludwig, *Prosperity Through Competition*. New York: Praeger Publishers, (c) 1958, p. 10.

62. Ibid., p. 14.

63. Samuelson, Paul A., "Road to 1984," *Newsweek*, August 13, 1973, p. 84. Copyright Newsweek, Inc. 1973, reprinted by permission.

64. "The World, Chile After Allende," *The Economist*, October 13, 1973, p. 43.

65. "People Even Get Tired of Prosperity," *U.S. News & World Report*, October 1, 1973., p. 72. "Bulletin of Labor Statistics," International Labor Office, Geneva, 2nd qtr., 1973.

66. "Big-Handed Palme," *The Economist*, August 25, 1973, p. 77.

67. Chamberlain, John, "Allendeism Appears in Canada, *Milwaukee Sentinel*, October 31, 1973, p. 16.

68. Skinner, op. cit., pp. 22-23.

69. Skinner, *Walden Two*. New York: Mac Millan Publishing Co., Inc., (c) 1948, pp. 296-7.

70. Orwell, George, *Nineteen Eighty-Four*. Harcourt Brace Jovanovich, Inc., (c) 1949), pp. 217-9.

71. Solzhenitsyn, Aleksandr, "A Soviet Martyr's Anguished Plea," *Wall Street Journal*, September 19, 1973. Reprinted with permission of The Wall Street Journal (c) 1973 Dow Jones & Company, Inc. All Rights Reserved.

72. Röpke, Wilhelm, *Economics of the Free Society*. Chicago: Henry Regnery Company, (c) 1963, p. 22.

73. From Organization for Economic Cooperation and Development Data.

74. Schumpeter, Joseph A., *Capitalism, Socialism & Democracy*. Scranton, Harper & Row, (c) 1949.

75. Dallas Times Herald, September 21, 1973.

76. Röpke, op. cit., pp. 253-5.

77. *Encyclopedia Britannica*, Vol. 7, op. cit., pp. 316-8.

78. White Lynn (Jr.), *Medieval Technology and Social Change*. Oxford University Press, (c) 1973, p. 39.

79. Ibid., p. 59.

80. Ibid., p. 68.

81. 1890 data—*Economic Growth in the United States*. New York: Committee for Economic Development, October 1969. 1930, 1950 data—"The National Income and Product Accounts of the United States, 1929-65 Statistical Tables," *Supplement to Survey of Current Business*, Department of Commerce, Bureau of Economic Analysis, Table 6.6, pp. 110, 112-13. 1972 data—*Survey of Current Business*, Table 6.6, July 1973.

82. "Agricultural Statistics," U. S. Department of Agriculture, 1972, p. 687, Table AO8.

83. "Survey of Current Business, "Department of Commerce, Bureau of Economic Analysis, July issues, Table 1.21.

84. "The National Income and Product Accounts of the United States, 1929-65." Statistical Tables, op. cit. Survey of Current Business, Table 6.6, July 1973.

85. Bell, Daniel, *The Coming of Post-Industrial Society: A Venture in Social Forecasting.* New York: Basic Books, (c) 1973, p. 147.

86. 1890 data—*Economic Growth in the United States,* op. cit. 1972 data—*Survey of Current Business,* Table 6.6, July 1973. Estimates beyond 1972 by Patrick E. Haggerty.

87. "Survey of Current Business," July 1973, Tables 1.10 and 6.19. Dividend tax liability estimated by Advanced Economic Planning, TI, using data from "Statistics of Income, 1970, Individual Income Tax Returns," Internal Revenue Service, Pub. #79 (10-72).

88. Ibid.

89. Public Opinion Poll, Opinion Research Corporation, Princeton, N.J., Spring 1972.

90. Ways, Max, "Business Needs to do a Better Job of Explaining Itself," *Fortune Magazine,* LXXXVI, #3 (September 1972), p. 86. Reprinted from the September 1972 issue of Fortune Magazine by special permission; (c) 1972 Time Inc.

91. *Science and Government Report,* November 1, 1973, p. 4.

92. "Counter-Attack—Business Acts to Improve a Tarnished Image," *U. S. News & World Report,* October 8, 1973, pp. 49-50.

93. Lerner, Max, Introduction to *Wealth of Nations* by Adam Smith. New York: Random House, (c) 1970.

94. Denison, Edward F., *Why Growth Rates Differ: Postwar Experience in Nine Western Countries.* Washington, D. C.: Brookings Institution, (c) 1967, Table 21-1, p. 298.

95. Ibid.

96. Solow, Robert from Paul Samuelson's *Economics,* 9th ed., op. cit., p. 748.

97. Denison, op. cit., Table 21-3, p. 300.

98. Department of Commerce; Office of Education, HEW.

99. Office of Education, HEW.

100. Bowen, William G., *The Economics of the Major Private Universities,* pp. 14-16. Reprinted with the permission of the Carnegie Commission on Higher Education. Copyright (c) 1968, William G. Bowen.

101. Friedman, Milton, *Capitalism & Freedom.* Chicago: University of Chicago Press, 1962, p. 89.

102. Department of Labor, Bureau of Labor Statistics.

103. *Monthly Labor Review,* June 1972, v. 95, #6, p. 30. U. S. Department of Labor, Bureau of Labor Statistics.

104. Holt, Charles C., et. al. *The Unemployment-Inflation Dilemma: A Manpower Solution.* Washington, D.C.: The Urban Institute, 1970, p. 35.

105. Ibid.

106. Ibid., p. 29.

107. Ibid.

108. Ibid., p. 36.

109. Friedman, Milton, *Dollars and Deficits,* Englewood Cliffs: Prentice Hall, (c) 1968, pp. 159-60.

110. "Employment and Earnings," V. 20, #4, October 1973, pp. 27, 29, 35, Tables A-1, A-3, A-8, A-9. U. S. Department of Labor, Bureau of Labor Statistics.

111. "Population Estimates and Projections," Series E, U. S. Dept. of Commerce, Bureau of the Census.

112. Coleman, James, "Youth—Transition to Adulthood," from "PSAC: Last Hurrah from Panel on Youth" by J. Walsh. *Science*, Vol. 182, pp. 141-5, 12 October 1973. Copyright 1973 by the American Association for the Advancement of Science.

113. "Further Preliminary Results of the New Jersey Graduated Work Incentive Experiment," Office of Economic Opportunity Pamphlet 3400-4, May 1971, p. 1.

114. Moynihan, Daniel P., *The Politics of A Guaranteed Income*, Random House, 1973.

115. OEO #3400-4, op. cit., p. 2.

116. Ibid., pp. 3-4.

117. Ibid.

118. Fried, op. cit., p. 168, 165.

119. "Work in America." Report of a special task force to the Secretary of Health, Education, and Welfare, Cambridge: MIT Press, 1973.

120. "Guide to Modern Profit Sharing," The Profit Sharing Council of America.

121. Ibid.

122. "Science Indicators 1972," Report of the National Science Board 1973. U.S. Govt. Printing Office, 1973, Fig. 14, p. 22.

123. Ibid., Fig. 15, p. 23.

124. Ibid., Fig. 18, p. 27.

125. Ibid., Fig. 1, p. 4.

126. Ibid., Fig. 2, p. 5.

127. Pyhrr, Peter A., *Zero Base Budgeting*. New York: John Wylie & Sons, 1973.

128. Mansfield, Edwin, "The Contribution of Research and Development to Economic Growth in the United States," *Papers and Proceedings of a Colloquium on Research and Development and Economic Growth/Productivity*. National Science Foundation. NSF 72-303. (Washington, D.C. 20402: Supt. of Documents, U.S. Govt. Printing Office), pp. 21-36.

129. "Science Indicators 1972," op. cit., p. 35.

130. *Research and Development and Economic Growth/Productivity* (Papers and Proceedings of a Colloquium). National Science Foundation. NSF 72-303.

131. Charnes, A., W. W. Cooper and G. Kozmetsky, "Measuring, Monitoring, and Modeling Quality of Life." *Management Science*. Providence: Institute of Management Sciences, Vol. 19, #10, June 1973.

132. Reich, op. cit., p. 348.